Programmed Cell Death

Edited by Hala Gali-Muhtasib
and Omar Nasser Rahal

Published in London, United Kingdom

IntechOpen

Supporting open minds since 2005

Programmed Cell Death
http://dx.doi.org/10.5772/intechopen.80192
Edited by Hala Gali-Muhtasib and Omar Nasser Rahal

Contributors
Carmen C. Diaconu, Petruta Gurban, Cristina Mambet, Mihaela Chivu-Economescu, Laura Georgiana Necula, Denisa Dragu, Saviana Nedeianu, Ana Iulia Neagu, Diana Cristodor, Coralia Bleotu, Lilia Matei, Aurelia Tatic, Mehtap Kara, Ezgi Öztaş, Hai Zhang, Zhinan Chen, Yibin Feng, Zhangfeng Zhong, Feiyu Chen, Ning Wang, Hor Yue Tan, Samo Ribaric, Hala Gali-Muhtasib

First published in London, United Kingdom, 2020 by IntechOpen
IntechOpen is the global imprint of INTECHOPEN LIMITED, registered in England and Wales, registration number: 11086078, 7th floor, 10 Lower Thames Street, London, EC3R 6AF, United Kingdom
Printed in Croatia

British Library Cataloguing-in-Publication Data
A catalogue record for this book is available from the British Library

Additional hard and PDF copies can be obtained from orders@intechopen.com

Programmed Cell Death
Edited by Hala Gali-Muhtasib and Omar Nasser Rahal
p. cm.
Print ISBN 978-1-78984-748-2
Online ISBN 978-1-78984-749-9
eBook (PDF) ISBN 978-1-83968-470-8

For EU product safety concerns:
IN TECH d.o.o., Prolaz Marije Krucifikse Kozulić 3, 51000 Rijeka, Croatia, info@intechopen.com or visit our website at intechopen.com.

Meet the editors

Hala Gali-Muhtasib is a tenured professor of cell and cancer biology at the American University of Beirut, Lebanon. She received her PhD from Kansas State University, Manhattan, USA. She is a prominent scholar with outstanding scientific contributions to the study of the role of natural products in cancer therapy, in addition identifying their cellular and molecular mechanisms of action. Her recent research interests include identifying stem cell therapeutics and enhancing drug delivery by nanoparticle encapsulation of natural products. She has published more than 95 peer-reviewed articles, more than 10 book chapters, and has edited a few books. Gali-Muhtasib has received many research awards for her contributions to advancing the field of natural product drug discovery and drug delivery.

Omar Nasser Rahal is a physician scientist who earned his MD degree with outstanding performance from SABA University School of Medicine, Caribbean, Netherlands. He earned his Bachelor of Science degree in Biology from the American University of Beirut, Lebanon, and was further appointed as a translational cancer researcher in the field of molecular biology and anticancer drug discovery. His research interests are in the field of oncology with a focus on translational research, systematic and comparative reviews, case studies, and clinical trials. Omar has co-authored six peer-reviewed articles and one book chapter. His passion for research along with his clinical career has contributed to the improvement of the quality of the healthcare system in the United States.

Contents

Preface

Programmed cell death is a normal and vital physiologic process, the aberrant regulation of which is a cancer hallmark and is a causative factor for many diseases. Knowledge of cell death mechanisms has expanded in the past years based on extensive research in the basic and clinical arena. This book compiles the latest research in the field and will be of interest to basic scientists, clinicians, and graduate students.

Hala Gali-Muhtasib
Department of Biology and Department of Anatomy,
Cell Biology and Physiological Sciences,
American University of Beirut,
Lebanon

Center for Drug Discovery,
American University of Beirut,
Lebanon

Omar Nasser Rahal
Beaumont Health Care System,
Dearborn, Michigan, USA

Section 1

Programmed Cell Death Mechanisms

Chapter 1

The Underlying Mechanisms of Chinese Herbal Medicine-Induced Apoptotic Cell Death in Human Cancer

Feiyu Chen, Zhangfeng Zhong, Hor Yue Tan, Ning Wang and Yibin Feng

Abstract

The high incidence of cancer is a global burden. Cancer cells acquire immortality, which results in loss of control in cell proliferation and population expansion. Cancer cells undergo a series of genomic instability, leading to mutated amplification or deletion of certain genes that strictly control the cell fate. Programmed cell death is a mechanism of cell fate control that is aberrantly regulated in cancer cells. Apoptosis is the major form of programmed cell death regulated by both intrinsic and extrinsic pathways. Discovering effective and specific alternative solutions that can reprogram apoptosis in cancer cells is always a challenge. Chinese herbal medicine has captured increasing attention from both researchers and manufacturers, as evidenced by observable curative effects from previous clinical experience. Hence, to clarify and reinforce the understanding of the effect of Chinese medicine on cancer, in this chapter, we will retrospectively review the latest 5 years of literature and summarize the mode of action of Chinese herbal medicine on apoptotic cell death in cancer. Both Chinese medicine-induced intrinsic and extrinsic mechanisms of apoptosis will be discussed, and common compounds from Chinese medicine with druggable potential as novel apoptosis-inducing agents will be highlighted.

Keywords: cell apoptosis, Chinese herbal medicine, programmed cell death, intrinsic and extrinsic pathways, human cancer

1. Introduction

Programmed cell death is a tight process mediated by an intracellular program, whereby damaged or harmful cells and their organelles are recycled or disposed of. The regulated cell death occurs along with morphological alterations. Biologists have employed such morphotypes from functional and biochemical perspectives to classify cell death routines. As provided by the Nomenclature Committee on Cell Death, three major widely accepted definitions of terms include apoptosis, autophagy, and necrosis [1]. Among them, apoptosis involves plasma membrane blebbing, cytoplasmic shrinkage, nuclear fragmentation, and chromatin condensation. The process ends up with formation of apoptotic bodies that are being taken up by phagocytic cells [1]. Apoptosis has been orchestrated in detail as two basic mechanism-oriented classifications: intrinsic pathway and extrinsic pathway.

Intrinsic apoptosis is initiated by multiple exogenous and endogenous stimuli such as stress, DNA damage, and chemotherapy, whereby cytochrome c is increasingly released from the mitochondria to the cytosol. Irreversible mitochondrial outer membrane permeabilization (MOMP) hence occurs, followed by caspase-3 activation that mainly promotes the typical apoptosis features [2]. In striking contrast, extrinsic apoptotic pathway is a modality of cell death driven by extracellular microenvironmental perturbations, which are normally two types of plasma membrane receptors: (1) death receptors consisting of a type of tumor necrosis factor (TNF) receptor FAS and TNF-related apoptosis-inducing ligand (TRAIL) receptors. The activation occurs when they bind to the cognate ligands; (2) dependence receptors such as NTN1, whose presence of elevated level relies on the drop in the levels of their specific ligands below a specific threshold. The process is then propagated by caspase-8 and executed mainly by caspase-3 [3] (**Figure 1**).

To our knowledge, programmed cell death provides an important function in metazoan cells to eliminate the toxic accumulation of superfluous cells and organelles and thereby to sustain the homeostatic cellular life. Due to loss of control of cancer cells in proliferation and expansion, a major focus has been placed on the association between defective regulation of apoptosis and immortality property of cancer cells. A large body of experimental evidence has unveiled that the defect in the physiological mechanism of apoptosis may promote tumorigenesis, while regulated apoptosis contributes to the recovery from diseases. For example, MOMP is one of the most common phenomena observed which is mediated by a variety of protein interactions of B-cell lymphoma 2 protein (Bcl-2) family, representing a total of 25 pro- and anti-apoptotic proteins. During the life of the cell, the balance of Bcl-2 family proteins partly determines cellular health. Besides the presence of decreased levels of pro-apoptotic proteins, the activation of anti-apoptotic family members is an important mechanism of apoptosis disequilibrium in cancer. Cellular inhibitors of apoptosis proteins (IAPs) represent a family of evolutionarily

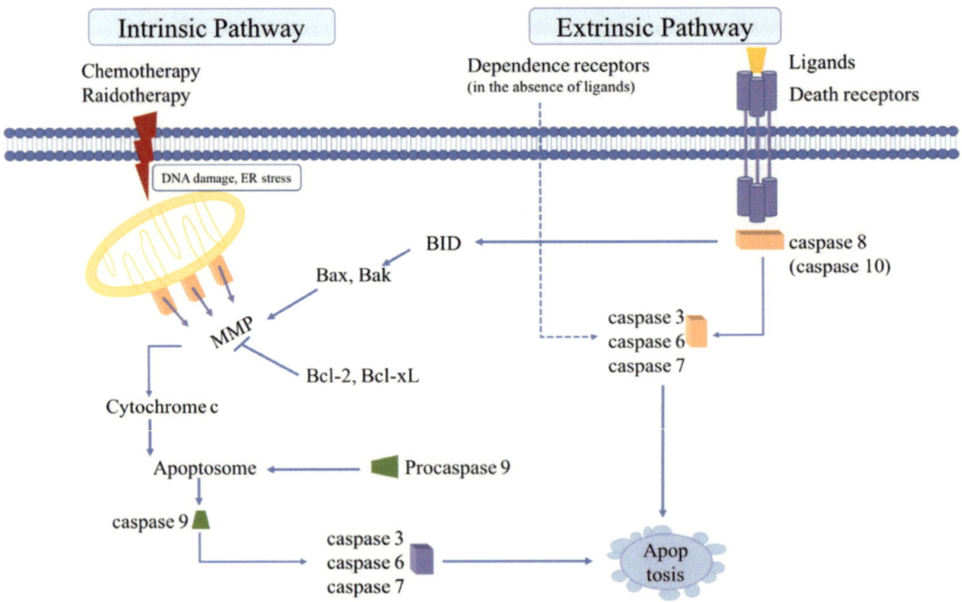

Figure 1.
Intrinsic and extrinsic pathways of apoptosis.

conserved apoptosis suppressors, which are known to be dysregulated in many cancers; that said, cancer cells might use the disequilibrium to retard apoptotic processes or stay in an apoptosis-resistant state [4, 5].

Therefore, seeking for sensitive and specific treatments that can restore the perturbation of apoptosis to a state of equilibrium in cancer cells is a challenge. Previous studies have focused on Chinese herbal medicine for its employment for centuries in the treatment of patients as well as its positive effects on tumors. Rather than focusing on the ablation of tumors per se, Chinese medicine focuses on correcting the imbalance of apoptosis. Evidence has been accumulating over decades that natural compounds of herbal medicine or formula are responsible for correcting apoptotic disequilibriums or resetting of apoptotic thresholds. A range of signaling pathways have been involved in the favorable effects of Chinese medicine in the treatment of cancers. As the domain continues to develop and novel molecular mechanisms are still being characterized, we have retrieved the latest 5 years of literature related to tumor apoptosis in response to herbal medicine. We also present literature on natural compounds from Chinese medicine with druggable potential as novel apoptosis-inducing agents.

2. Chinese herbal medicine that induces intrinsic apoptosis in human cancers

The intrinsic apoptosis pathway is driven by intercellular and extracellular perturbations such as oxidative stress, DNA damage, and chemo- and radiotherapies, which result in mitochondrial dysfunction and release of cytochrome c in the cytoplasm. Initiator caspase-9 is then activated, and executioner caspase-3 precipitates the apoptotic process. During this process, Bcl-2 family proteins which are involved in the activation of intrinsic apoptosis have been identified to play a role by either activating pro-apoptotic pathways that cause the subsequent efflux of cytochrome c or inhibit cytochrome c release. Pro-apoptotic proteins include Bax, Bid, and Bad, among others. The other subtype presenting anti-apoptotic activity contains Bcl-2 and Bcl-xL. When procaspase-9 forms, the caspase cascade will be in turn activated and caspase-2, caspase-8, caspase-9, and caspase-10 initiate the process of apoptosis, while caspase-6, caspase-7, and mainly caspase-3 precipitate the cell apoptosis. This is totally different from extrinsic apoptosis, which is mediated by death receptors. As membrane receptors such as Fas, TNF receptors will interact with corresponding ligands to recruit relevant adaptor proteins, followed by the recruitment of a series of downstream factors, in particular caspase-8, which is the critical mediator to activate the caspase cascade [6, 7] (**Figure 1**).

Evidence has been accumulating, across the last few decades, that dysfunctional apoptosis in cancer partly leads to the immortality property of cancer cells [6]. With well-reported observations of good curative effects on multiple cancers as well as clinical application of centuries, attention has been extensively attracted to Chinese herbal medicine. Substantial laboratory evidence has unveiled that Chinese herbal medicine is able to recover the defective apoptosis via intrinsic apoptotic pathway, which eventually is in favor of tumor suppression (**Table 1**). Interestingly, Chinese herbal medicine that only involves the regulation of the extrinsic pathway has been rarely reported.

Osteosarcoma is a type of cancer existing in a bone. Most cases occur among children and adolescents. Polyphyllin I (PPI) is extracted from *Paris polyphylla* rhizomes, which has been used for centuries in China for the treatment of infectious diseases and cancer. In osteosarcoma cells, PPI has been shown to activate unfolded protein response (UPR)/endoplasmic reticulum (ER) stress pathway, followed by

Name of Chinese herbal medicine	Cancer type	Mechanism of action	Ref.
Paris polyphylla Smith var. yunnanensis (polyphyllin I)	Osteosarcoma	UPR and ER stress	[8, 9]
Salvia miltiorrhiza (tanshinone IIA)	Osteosarcoma	Bcl-2 regulation	[10]
	Cervical cancer	ER stress	[11]
Epimedium (icaritin)	Cervical cancer	ROS stress	[12]
Chinese bayberry (flavonoids)	Ovarian cancer	Bcl-2 regulation	[13]
Psoralea corylifolia	Breast cancer	Bcl-2 regulation	[14]
Pueraria (6″-O-xyloside)	Colon cancer	Bcl-2 regulation	[16]
Ornithogalum caudatum Ait (OSW-1)	Colon cancer	Cytochrome c release	[17]
Ginkgo biloba (ginkgolic acids)	Colon cancer	ROS stress	[18]
Macleaya cordata (sanguinarine)	Colorectal cancer	Bax regulation	[20]
Cordyceps sinensis (cordycepin)	Pancreatic cancer	Bcl-2 regulation	[22]
Angelica sinensis (N-butylidenephthalide)	Gastric cancer	MMP	[23]
Lonicera japonica (luteolin)	Gastric cancer	Bax, Bcl-2 regulation	[24]
Tripterygium wilfordii (celastrol)	Liver cancer	ER stress	[27]
	HCC	Bax, Bcl-2 regulation	[28]
Polygoni multiflora (ethanolic extract)	HCC	MMP	[29]
Brucea javanica (dehydrobruceine B)	Lung cancer	MMP	[30]
Catharanthus roseus (cathachunine)	Leukemia	ROS stress	[31]
Pulsatilla chinensis (saponins)	Leukemia	MMP	[32]
Coptidis rhizoma (berberine)	Melanoma	Bax, Bcl-2 regulation	[33]
	HCC	Cytochrome c release	[34]

UPR, unfolded protein response; ER, endoplasmic reticulum; Bcl-2, B-cell lymphoma 2 protein; ROS, reactive oxygen species; Bax, Bcl-2-associated X protein; MMP, mitochondrial membrane potential; HCC, hepatocellular carcinoma.

Table 1.
Chinese herbal medicines that induce intrinsic apoptosis in human cancer.

descending levels of anti-apoptotic proteins as well as ascending expressions of pro-apoptotic proteins [8, 9]. With regard to osteosarcoma, tanshinone IIA (Tan IIA) is one of the main phytochemical ingredients isolated from the roots of *Salvia miltiorrhiza* (Danshen). Exposure of osteosarcoma cells to Tan IIA caused in vivo tumor suppression and apoptosis induction in osteosarcoma cells. Huang et al. investigated the mechanism of its inhibitory effect and found that Tan IIA treatment elicited significant activation of caspase cascade by Bcl-2 family modulation, as evidenced by a remarkable increase in the fission protein Drp1 and a decrease in mitochondrial fusion proteins Mfn1/2 and Opa1. The study concluded that mitochondrial dysfunction in combination with dynamic change was involved in apoptosis of primary malignant bone tumors treated by Tan IIA [10]. More so, Tan IIA exhibited strong inhibitory effects on cervical carcinoma CaSki cells through promoting caspase cascades, whereas the phosphorylation of p38 and JNK signaling was activated. Comprehensive proteomics revealed the global protein changes and the network analysis confirming that Tan IIA administration activated ER stress signaling cascade that eventually resulted in mitochondrial-related apoptosis [11].

Cervical cancer is the fourth most common female malignancy in the world. Despite preventive vaccines against human papillomavirus (HPV) which are now commercially available and which have been shown to be safe and effective, there are still a large number of women, in particular in low- and middle-income countries, who are less likely to have access to HPV vaccines or screening of cervical cancer due to geographical, economic, and political barriers. Icaritin, a native compound derived from the Chinese herb *Epimedium*, was demonstrated to be effective in repressing growth of human cervical cancer cells such as HeLa and SiHa. The levels of pro-apoptotic protein Bax and activated caspase-3 and caspase-9 enzymes were upregulated, with concomitant downregulated levels of anti-apoptotic proteins Bcl-2 and XIAP. The changeable expression of these proteins had implications in remarkable induction of apoptosis in icaritin-treated cancer cells, which suggested that cancer cell death via induction of extensive oxidative DNA damage was promoted by icaritin-induced ROS overload that rendered activation of the intrinsic apoptotic pathway [12].

Ovarian cancer and breast cancer are malignancies that commonly occur in females. The treatment with Chinese bayberry leaf flavonoids increased the expression of cleaved caspase-3 and caspase-7, which induced intrinsic apoptosis with the activation of Erk-dependent caspase-9, as well as increased expression of the pro-apoptotic proteins Bad and Bax and decreased levels of anti-apoptotic proteins Bcl-xL and Bcl-2 [13]. Bakuchiol is an active constituent of Chinese herb *Psoralea corylifolia* which induced disturbed mitochondrial membrane potential in MCF-7 cells. Bakuchiol-induced apoptosis was associated with increased expression of caspase family and Bcl-2 family proteins, suggesting that bakuchiol may induce apoptosis via the intrinsic apoptotic pathway [14]. Another study evaluated the function of *Paeonia suffruticosa* in triple-negative breast cancer cells. Bcl-2 expression was found to be decreased, while Bax levels remained relatively constant. Small decrease in Fas ligand levels was observed in parallel with a lack of increase in caspase-8 activity. The extract was able to induce intrinsic apoptosis which meant it possessed the potential ability of reducing cancer burden [15].

Colon cancer is worldwide and is considered the third most commonly diagnosed cancer clinically. To identify novel specific and effective therapeutic strategies for colon cancer is extremely essential. Natural products have gained increasing attention lately. Puerarin 6″-O-xyloside (PRX), a natural compound, is derived primarily from the root of the *Pueraria*. PRX was found to significantly upregulate cleaved caspase-3, cleaved caspase-9, Bcl-2, Bcl-2-associated X proteins, and phosphorylated c-Jun terminal kinase and downregulate expression levels of matrix metalloproteinase-3, metalloproteinase-9, and vascular endothelial growth factor. The study suggested that PRX exerted antitumor activity against colon cancer cell lines and the anticancer mechanisms of PRX may be associated with the induction of mitochondria-mediated intrinsic apoptosis, which provides a scientific basis for the clinical use of PRX in the treatment of colon cancer [16]. As a natural compound, *Ornithogalum caudatum* Ait is primarily used as an anti-inflammatory and antitumor agent in Chinese folk medicine. It was shown that with low toxicity on normal colonic cells, an isolated compound OSW-1 suppressed colon cancer cells in vitro via intrinsic apoptotic pathway, whereby it increased cellular calcium, changed mitochondrial membrane potential, disrupted mitochondrial morphology, and led to the release of cytochrome c and the activation of caspase-3 [17]. Ginkgolic acids (GA), a botanical drug extracted from the seed coat of *Ginkgo biloba* L., possess various bioactive properties. The findings, for the first time, illustrated that GA suppressed colon cancer cell proliferation, migration, and invasion ability. Ginkgolic acids (GA) proved to trigger intrinsic apoptosis as evidenced by the release of cytochrome c. Autophagy modulation mediated by ROS generation was also observed in GA-treated colon cancer cells, elucidating that GA might be a potential agent for colon cancer therapy [18].

Colorectal cancer and colon cancer are clearly related and often used interchangeably. These two terms are often believed to be the subset of the other or even the same thing. In truth, despite similarities, there is still variation including sex predilection, anatomy, disease recurrence, surgery, and invasion of nearby tissues, not the least of which are the development ways of the two diseases. *Macleaya cordata* is originally described in Ben Cao Shi Yi in Tang dynasty and is commonly used in the treatment of various diseases for thousands of years [19]. Sanguinarine is a major bioactive component of *Macleaya cordata*. Sanguinarine was shown to decrease the tumor size of implanted colorectal BALB/c-nu mice model via the intrinsic apoptosis pathway with significant increased cleavage of caspase-3 and poly(ADP-ribose) polymerase (PARP) in orthotopical colorectal carcinoma. In vitro experiments found that sanguinarine could increase mitochondrial ROS and trigger mitochondrial membrane potential (MMP) in multiple colorectal cancer cell lines. Furthermore, intrinsic apoptosis induced by sanguinarine was demonstrated to be Bax-dependent [20]. Cordycepin is one of the main native constituents extracted from the traditional Chinese herbal remedy *Cordyceps sinensis* and *Cordyceps militaris*. It has been extensively used as food, health supplement, and herbal formulas from ancient times for health care [21]. Recent evidence demonstrated that cordycepin has anti-inflammatory and antitumor activities. In human MIA PaCa-2 and Capan-1 pancreatic cancer cells, cordycepin was found to inhibit cell viability, proliferation, and colony formation ability and induce cell cycle arrest and early apoptosis in a dose- and time-dependent manner, while the same effect was observed in in vivo experiments. Further, the expression levels of Bax, cleaved caspase-3, cleaved caspase-9, and cleaved PARP were upregulated, and Bcl-2 proteins were downregulated. The study suggested that either in vivo or in vitro the intrinsic apoptotic pathway mediated by mitochondria was involved in the cordycepin's antitumor capacity [22].

Despite the fact that the incidence of gastric cancer has declined across the globe over the past century, there is still a startling lack of effective therapeutics. *Angelica sinensis* (Danggui) is one of the most famous medical herbs widely used in China. Liao et al. investigated the function of a bioactive compound N-butylidenephthalide (BP) from Danggui in gastric cancer. The results showed that BP inhibited gastric cancer cell proliferation and induced apoptosis through activating mitochondrial apoptotic pathway. These data provide the basis for a novel therapeutic approach toward the management of gastric cancer [23]. Another active component named luteolin from a traditional Chinese medicine exhibits potent antitumor properties. The molecular events occurring in the process of tumor inhibition and the signal transduction pathways involved were explored by Lu et al. They found increasing levels of caspase-3, caspase-9, and cytochrome c in response to luteolin as well as an increased ratio of Bax to Bcl-2, suggesting that luteolin induced apoptosis through the intrinsic pathway [24].

Globally, liver cancer ranks as the sixth most common form of cancer. With the growing prevalence of liver cancer, it remains a major killer worldwide. Patients' status is dismayingly unsatisfying unless liver cancer is caught early and specific therapeutic methods are being discovered [25]. Celastrol is a pharmacologically active compound originally identified from the root bark of the Chinese herb "Thunder of God Vine" (*Tripterygium wilfordii* and *Celastrus regelii*). Since old times it has been used as a natural remedy for inflammatory conditions and autoimmune diseases [26]. Investigators demonstrated the inhibitory effects of celastrol on liver cancer HepG2 and Bel7402 cell lines. Induction of ER stress and UPR occurred in cells exposed to celastrol, which subsequently activated the intrinsic apoptotic pathway. They also reported that celastrol repressed H22 tumor growth in mice model via ER stress induction [27]. Another report aimed to evaluate the antitumor effects of celastrol against diethylnitrosamine (DEN)-induced hepatocellular carcinoma (HCC) in rats. In addition, the underlying mechanism was explored, and the data showed that celastrol activated the intrinsic mitochondrial apoptosis pathway, inhibited anti-apoptotic Bcl-2 and Bcl-xL, and induced pro-apoptotic Bax, cytochrome c, PARP, and caspases [28]. Zhiheshouwu (*Polygoni Multiflori Radix Praeparata*) is a Chinese medicinal herb exhibiting inhibitory effects on cancer cells. The study investigated the function of Zhiheshouwu ethanolic extract (HSWE) and revealed the decreased mitochondrial membrane potential in HSWE-treated Bel-7402 cells. The authors concluded that HSWE induced intrinsic apoptosis in hepatocellular carcinoma cells on the basis of the evidence that mitochondrial injury is characterized as an intrinsic apoptotic cell death mechanism [29].

Lung cancer has been one of the leading causes of mortality in this era. With the etiologic factors of lung cancer being more complex such as environmental pollution, industrialization, and urbanization, cases of lung cancer increase across the world and account for nearly 20% of cancer-caused deaths. Studies on the new ways of diagnosis and treatment have played an important role in the tertiary prevention of lung cancer. *Brucea javanica* is an effective traditional medicine listed in Chinese Pharmacopeia. It has long been used as a commercially available agent for cancer treatment in practice. Dehydrobruceine B (DHB) is an active ingredient isolated from *Brucea javanica*. Since the loss of MMP, the release of cytochrome c into cytosol and the cleavage of caspase-9, caspase-3, and PARP were observed in lung cancer cells exposed to DHB. Researchers suggested that DHB-induced apoptosis was mediated through mitochondrial intrinsic pathway [30].

Leukemia is a group of life-threatening malignant disorders which often originate in the blood and bone marrow. The acute leukemia is more prevalent among adolescent and young adult population. *Catharanthus roseus*, a species of flowering plants, consists of dimeric indole alkaloids with significant antitumor activities. The induction of apoptosis by cathachunine occurred along with the regulation of Bcl-2 protein family members. The observations further indicated that cathachunine triggered ROS-dependent mitochondria-mediated intrinsic pathway in human HL60 and K562 leukemia cells, which provided evidence for a natural source of an antitumor agent [31].

Total saponins isolated from *Pulsatilla chinensis* has been identified to induce the apoptosis of solid cancer cells. The rhizoma of the plant has virtually been used as Chinese herbal remedy for thousands of years. 23-Hydroxybetulinic acid, one of natural compounds from total saponins, upregulated Bax, cytochrome c, cleaved caspase-9, and caspase-3 expressions and downregulated Bcl-2 and survivin levels, suggesting that saponins induced intrinsic apoptosis via disrupting mitochondrial membrane potential [32].

Coptidis rhizoma (CR) has been used in clinical practice from thousands of years ago. A large body of research has placed attention on *Coptidis rhizoma* as well as its extracts and major active chemical constituents. Berberine, the most famous natural ingredient from CR, has been shown to have anticancer activities against multiple cancers. Recent work investigated the effects of berberine in human melanoma and reported that exposure of CR to human melanoma cells triggered significant suppression of anti-apoptotic proteins including Bcl-2, Mcl-1, and Bcl-w while upregulating the expression of pro-apoptotic proteins Bax and Bak [33]. As berberine is one of the most active molecules from traditional Chinese medicine, to better understand the natural product, our group has conducted several investigations over the years. Although berberine-induced cell death has been extensively demonstrated in cancer, the underlying death mechanisms still remain obscure. Exposure of hepatocellular carcinoma HepG2 and MHCC97-L cells to berberine increased Bax expression, permeable transition pore formation, cytochrome c release to cytosol, and subsequent execution of the caspase-3 and caspase-9 [34]. Tumor suppressor p53 plays an important role in cancer inhibition. It was verified to be involved in berberine's antitumor action [35, 36]. Our group found that human HCC cell miR-23a was upregulated upon berberine treatment, and the upregulation of miR-23a could be blocked by inhibiting p53 expression. The study suggested that miR-23a may be involved in regulating the antitumor effect of berberine in HCC through p53-dependent mechanisms [37]. Baicalin is a natural flavonoid from several medicinal herbs such as *Scutellaria baicalensis* Georgi. We speculated that tumor-associated macrophages (TAM) had a key role in HCC. Our findings revealed that TAM repolarisation contributed to suppressive function of baicalin on HCC and autophagy-associated activation of RelB/p52 was essential in the process [38]. Huanglianjiedu decoction (HLJDD) has been well documented for the treatment of heat and dampness-related diseases thousands of years ago. As clinical practice requires more specific and safe Chinese herbal formula, our group members explored the inhibitory effect of this formula in HCC suppression. The results showed the involvement of eEF2 inhibition in its mode of action [39].

3. Both intrinsic and extrinsic apoptotic pathways involved in Chinese herbal medicine-induced apoptosis in human cancers

From a molecular standpoint, extrinsic apoptosis and intrinsic apoptosis are strikingly different. But on the whole, the two pathways are sometimes related. Receptor trimerization can lead to recruitments of several death domains and subsequent recruitment and activation of caspase-8 and caspase-10, which then either activate the intrinsic apoptotic pathway through cleavage of the BID or initiate extrinsic apoptosis directly by activating executioner caspase-3, caspase-6, and caspase-7 to induce efficient cell death (**Figure 1**). Previous efforts have reported that both intrinsic and extrinsic apoptoses may coexist when cancer cells were exposed to Chinese herbal medicine (**Table 2**).

Butein is a subtype of chalcones, which is widely biosynthesized in plants. Butein has been identified to be extractable from Chinese herbal medicine and possess different pharmacological activities. Recent works reported that butein owned abilities of inhibiting proliferation and inducing apoptosis both in vivo and in vitro. A finding suggested that butein could decrease cervical cancer cell viability via pro-apoptotic effect, which involved inhibition of IAP proteins and activation of both extrinsic and intrinsic pro-apoptotic pathways. Therefore, butein may be applicable for cervical cancer treatment [40]. *Tetrastigma hemsleyanum*, named Sanyeqing in Chinese, has long been used as a folk medicine to overcome cancer. A recent study prepared petroleum ether fractions (PEF) of Sanyeqing and aimed to investigate the possible mechanisms by which PEF presented antitumor activity against HeLa cells. Caspase-9 and caspase-3 were activated, and mitochondrial membrane potential decreased after PEF treatment. In addition, PEF administration was involved in the extrinsic apoptotic pathway indicated by the activation of caspase-8 [41]. The same observations were found in ovarian cancer cells upon exposure to butein. Increased levels of cytochrome c, caspase-3, caspase-8, and caspase-9 in two types of ovarian cancer cells, with concomitant downregulated levels of Bcl-2 and Bcl-xL and upregulated proteins Bax and Bad, suggested that both extrinsic and intrinsic pathways were involved in butein-induced apoptosis [42]. To treat ovarian cancer,

Name of Chinese herbal medicine (active constituent)	Cancer type	Mechanism of action	Ref.
Dalbergia odorifera, Caragana jubata, Rhus verniciflua, Semecarpus anacardium (butein)	Cervical cancer	IAP; Fas	[40]
Tetrastigma hemsleyanum (petroleum ether fraction)	Cervical cancer	MMP; Fas	[41]
Dalbergia odorifera, Caragana jubata, Rhus verniciflua, Semecarpus anacardium (butein)	Ovarian cancer	Bcl-2, Bcl-xL; Fas	[42]
Chinese bayberry (prodelphinidins)	Ovarian cancer	Bcl-2, Bcl-xL; DR5, Fas	[43]
Sanguisorbae (ziyuglycoside I)	Breast cancer	Bax, Bcl-2; Fas	[44]
Paeonia moutan (paeonol)	Prostate cancer	Bax, Bcl-2, caspase-9; caspase-8	[45]
Ginkgo biloba (exocarp extracts)	Lewis lung cancer	Bax, Bcl-2; Fas, p-p38	[46]
Labiatae (*Scutellaria barbata* D)	Lung cancer	Cytochrome c; Fas	[47]
Anisomeles indica (ovatodiolide)	Lung cancer	PUMA, Bax; DR5	[48]
Curcuma (curcuminoids)	Lung cancer	MMP, Fas	[49]
	Head and neck squamous cell	Caspase-9; caspase-8	[50]
	Leukemia	Bcl-2; FasL	[51]
Tripterygium wilfordii (celastrol)	Liver cancer	Caspase-9; caspase-8	[52]
	Mantle cell lymphoma	Caspase-9; caspase-8	[53]
Licorice (licochalcone B)	Oral squamous cell carcinoma	ROS; DR4	[54]
	Skin cancer	Sp1; Fas	[55]
Magnolia officinalis (honokiol)	Glioblastoma	Bax, Bcl-2, caspase-9; caspase-8, Fas	[56]
Allium sativum (allicin)	Glioblastoma	MMP; Fas	[57]
Phellinus linteus (hispolon)	Leukemia	Bax, Bcl-2, caspase-9; caspase-8	[58]
Jujube (extracts of jujube seed)	Leukemia	Caspase-9; caspase-8	[59]
Cordyceps kyushuensis Kob (aqueous extracts)	Leukemia	Bcl-2; Fas	[60]

IAP, inhibitor of apoptosis; MMP, mitochondrial membrane potential; Bcl-2, B-cell lymphoma 2 protein; Bcl-xL, B-cell lymphoma-extra large; DR5, death receptor 5; PUMA, p53-upregulated modulator of apoptosis; FasL, Fas ligand; DR4, death receptor 4; Sp1, specificity protein 1; Bax, Bcl-2-associated X protein.

Table 2.
Chinese herbal medicines that induce both intrinsic and extrinsic apoptosis in human cancer.

increasing herbal medicines have been studied. Chinese bayberry leaves are rich in prodelphinidins. Since the isolation and purification of prodelphinidins are difficult, the association between the degree of prodelphinidin polymerization and their anti-carcinogenic activity remains ambiguous. Recent findings reported that apoptosis was executed through the intrinsic pathway by upregulating the expression of pro-apoptotic proteins including p53-upregulated modulator of apoptosis (PUMA), Bax, and Bcl-2-associated agonists of cell death. The extrinsic pathway was also observed in the apoptotic process as evidenced by upregulation of death receptor 5 (DR5) and Fas expression [43]. Triple-negative breast cancer (TNBC) is currently considered as one of the most severe malignancies due to poor prognosis and aggressive clinical behavior. Recent work explored the cytotoxic effect of ziyuglycoside I, the major component extracted from *Sanguisorbae Radix*, on a type of TNBC cell line. The results showed that ziyuglycoside I triggered apoptosis of MDA-MB-231 cells in a dose-dependent fashion, as evaluated by the elevated expression of p53 and p21 and upregulated Bax/Bcl-2 ratio, suggesting that these effects were mediated through mitochondrial-initiated and Fas/FasL-initiated apoptotic pathways [44]. Paeonol (Pae) is a main active ingredient from the root bark of *Paeonia moutan*. Numerous reports indicated that Pae effectively restrained several types of cancer lines. A study lately reported that Pae inhibited cancer cell proliferation in prostate cancer. Moreover, the study showed that antiproliferative effects of Pae may be closely related to the activation of extrinsic and intrinsic apoptotic pathways [45].

Ginkgo biloba is known as an edible traditional Chinese medicine. The extracts prepared from the exocarp of *Ginkgo biloba* (GBEE) have been identified to possess capabilities of antitumor, antiaging, and immune promotion, among others. Lewis lung cancer cells were used as a cell model to detect the effect of GBEE, and results showed that Bax/Bcl-2 ratio and the release of cytochrome c from the mitochondria to cytosol increased. In addition, GBEE upregulated the cleaved caspase-3 protein expression as well as the protein levels of Fas, FasL, and p-p38. These data suggested that GBEE induced apoptosis in Lewis lung cancer cells via death receptor-mediated extrinsic pathways and mitochondrial-mediated intrinsic pathways [46]. *Scutellaria barbata* D. Don (SB) is a well-known anti-inflammatory compound isolated from dried whole plant of *Labiatae*. Chen et al. reported that the antitumor mechanism of SB was mediated by P38/SIRT1-regulated cell apoptosis, which involved mitochondrial- and Fas/FasL-mediated pathways [47]. With regard to lung cancer, ovatodiolide, extracted from medicinal herb *Anisomeles indica*, was also explored due to its effective antibacterial and anti-inflammatory properties. Recent work investigated the antitumor activity of ovatodiolide, in which the mechanism was characterized by elevated levels of PUMA, Bax, and DR5 proteins, decreased expressions of Bcl-2 and Mcl-1, as well as activation of caspase-8, caspase-9, and caspase-3 [48]. Curcuminoids, a mixture of curcumin, demethoxycurcumin (DMC), and bisdemethoxycurcumin (BDMC), are a primarily natural phenolic compound purified from *Curcuma* species. The study found that BDMC decreased the levels of MMP and promoted caspase-3, caspase-8, and caspase-9 activation while upregulating the levels of Fas ligand and Fas. These results suggested that BDMC triggered cell apoptosis via extrinsic and intrinsic signaling pathways [49]. Another group also examined the anticancer activity of curcumin against head and neck squamous cell. Treatment of FaDu and Cal27 cells with curcumin increased caspase-9 and caspase-8 protein expression, which meant that intrinsic and extrinsic apoptotic pathways were both activated by curcumin in the treatment of head and neck squamous cell carcinoma [50]. Besides, curcumin function was also investigated in leukemia cells. The results of PCR and Western blotting analysis showed that curcumin increased the FasL level; inhibited Bcl-2, NF-κB, and ERK expression; and activated P38 MAPK, JNK, and caspase-3. This study demonstrated that curcumin played its role not only through intrinsic but also through extrinsic apoptosis pathways [51].

Celastrol is an active ingredient derived from *Tripterygium Wilfordii*, a type of traditional Chinese medicinal herb, which has been reviewed for the treatment of liver cancer as stated above. As of now, celastrol has gained great interest due to its potential anti-inflammatory and antitumor properties in many cancers. For instance, the long-term survival of osteosarcoma has stagnated over decades; exposure of osteosarcoma to celastrol resulted in the activation of caspase-3, caspase-8, and caspase-9, which suggested that celastrol-induced apoptosis was mediated by extrinsic and intrinsic pathways [52]. Another report demonstrated that celastrol treatment markedly inhibited mantle cell lymphoma cells proliferation by stimulating apoptosis via extrinsic and intrinsic pathways while exerting minimal cytotoxic effects on normal cells. The results provided support for the clinical use of celastrol [53]. Licochalcone B (Lico B) belongs to retrochalcone family and is normally isolated from the roots of Chinese *licorice*, which has long been used in China with a variety of pharmacological properties such as antioxidant, anti-inflammatory, and antibacterial. The underlying mechanism by which Lico B plays a part in oral squamous cell carcinoma (OSCC) has been elucidated by Oh et al. They reported that exposure of Lico B to oral cancer cells induced upregulation of Bax as well as downregulation of Bid, Bcl-xL, and Mcl-1. The loss of MMP led to the release of cytochrome c. Furthermore, Lico B promoted the generation of ROS, which in turn induced DR4, DR5, and CHOP. Thereby, it is suggested that Lico B triggered apoptotic cell death via intrinsic and extrinsic pathways [54]. Lico B was also investigated in human skin cancer cells. Lico B induced apoptotic cell death through the modulation of specificity protein 1 and apoptotic proteins including death receptors, critical factors of the extrinsic pathway. Based on these facts, conclusions could be made that Lico B treatment resulted in extrinsic and intrinsic apoptotic cell death [55]. A natural bioactive compound honokiol that was isolated from the *Magnolia officinalis* exhibited potent inhibitory activity against multiple human cancer cells. Zhang et al. for the first time reported that the antineoplastic effect of honokiol on glioblastoma cells was mediated through caspase-dependent apoptosis that involved intrinsic and extrinsic signaling pathways [56]. Following treatment with allicin, the expression levels of Fas/FasL increased and Bcl-2 protein significantly decreased in glioblastoma cells, at both mRNA and protein levels. The data demonstrated that allicin induced glioma cell apoptosis by both extrinsic Fas/FasL-mediated and intrinsic mitochondrial pathways [57]. Hispolon was extracted from *Phellinus linteus* and has been demonstrated to show strong anticancer, anti-inflammatory, antioxidant activities. A study reported that exposure of human NB4 leukemia cells to hispolon resulted in upregulated expressions of apoptosis-related proteins, including the cleavage form of caspase-3, caspase-8, and caspase-9, the increased ratio of Bax/Bcl-2, and cytochrome c, with concomitant increased levels of Fas and FasL. Therefore, it was demonstrated that both extrinsic and intrinsic apoptotic pathways were involved in human leukemia cells treated with hispolon [58]. *Jujube* (Zǎo) is well known as a type of snack and has long been used as a supplement in gynaecopathia. Seeds of *jujube* exhibit antineoplastic effects and have been used in Chinese medicine for centuries. Recent work found that extracts of *jujube* seed could increase caspase-8 and caspase-9 activities in human Jurkat leukemia T cells through extrinsic and intrinsic apoptosis pathways [59]. Zhao and colleagues measured the antitumor activity of aqueous extracts of *Cordyceps kyushuensis* Kob (AECK), which is a type of entomogenous fungi. The group reported the upregulated amount of Ca^{2+} and downregulated expression of Bcl-2, which indicated that AECK triggered intrinsic apoptosis. Meanwhile, AECK gave rise to extrinsic apoptosis via elevating the level of Fas death receptor in U937 cells [60].

4. Chinese herbal medicine that is favorable in reducing the drug resistance in human cancers

Despite significant improvements in cancer treatment and emergence of a substantial number of novel therapeutics, cure rates for most malignancies remain suboptimal. Treatment resistance is less likely predicted for individual patients and is being the largest obstacle to the success of recovery. The most targeted therapies and chemotherapeutics for cancer disrupt cancer cells via the generation of pro-death signaling molecules and subsequent initiation of programmed cell death. Based on facts mentioned previously, defects in apoptotic pathways that make tumor cells fail to die are believed to be one of the reasons for resistance acquisition. Successfully targeting apoptotic pathways with Chinese herbal medicines may shed a new light on cancer therapy (**Table 3**).

Pterostilbene is a natural polyphenolic compound chemically related to resveratrol, which has received FDA GRAS status in 2007. Pterostilbene exhibits antitumor, antioxidant, and anti-inflammatory activities and is primarily found in blueberries, almonds, grape leaves, and vines. Effects of pterostilbene on cisplatin-resistant oral cancer cells and the mode of action were explored by researchers. By using pan-caspase inhibitor and directly testing DNA breakage of human oral CAR cells, pterostilbene was found to trigger caspase-dependent apoptosis, suggesting that intrinsic apoptotic cascade was involved in the effect of pterostilbene in oral cancer [61]. A heat-sensitive sesquiterpene named furanodiene is extractable from the essential oil of *Rhizoma Curcumae*. In doxorubicin-resistant MCF-7 cells, furanodiene was identified to preferentially cause apoptotic cell death by interfering with abnormal intrinsic- and extrinsic-dependent pathways [62]. As depicted in the introduction part, TRAIL has a specific antineoplastic property against malignancies. You and colleagues studied the function of trichosanthin, a kind of traditional Chinese medicine isolated from the root of *Trichosanthes*, on TRAIL resistance by using non-small cell lung cancer TRAIL-resistant cells. The results showed that the expression levels of extrinsic and intrinsic apoptotic proteins were modulated. They concluded that trichosanthin rendered apoptosis by augmenting the sensitivity of TRAIL-resistant cells through upregulating DR4 and DR5 [63]. Oxaliplatin is an effective alternative treatment of HCC after sorafenib treatment failure. The combination treatment of oxaliplatin and huaier was verified to exhibit remarkable synergistic antineoplastic effect through inhibition of expression of apoptosis-related proteins and Yes-associated protein (YAP), which was demonstrated to reduce the chemotherapeutic sensitivity of oxaliplatin. As such, huaier was considered to enhance the oxaliplatin sensitivity by modulating Yap as well as apoptotic processes [64]. Chinese people have used milky sap or the aboveground part of *Euphorbia lunulata* to treat cancerous ailments since old times. The specific

Name of Chinese herbal medicine (active constituent)	Cancer type	Resistant drugs	Ref.
Blueberry (pterostilbene)	Oral cancer	Cisplatin	[61]
Curcumae (furanodiene)	Breast cancer	Doxorubicin	[62]
Trichosanthes (trichosanthin)	Non-small cell lung cancer	TRAIL	[63]
Huaier (extracts)	Hepatocellular carcinoma	Oxaliplatin	[64]
Euphorbia lunulate (extracts)	Gastric cancer	TRAIL	[65]
Tanshinone (tanshinone IIA)	Gastric	Doxorubicin	[66]
Wild mushroom (clitocine)	Colon cancer	TRAIL	[67]
Curcuma longa (curcumin)	Oral cancer	Cisplatin	[68]
Tripterygium wilfordii (celastrol)	Oral cancer	Vincristine	[69]
Houpo (honokiol)	Glioma cancer	Temozolomide	[70]

TRAIL, tumor necrosis factor-related apoptosis-inducing ligand.

Table 3.
Chinese herbal medicines that reduce drug resistance in human cancer.

mode of action remains obscure and was lately elucidated by researchers. Fu et al. observed that *Euphorbia lunulata* extract significantly increased activities of apoptotic indexes including caspase-3, caspase-8, caspase-9, and Bax while downregulating Bcl-2 expression, suggesting that both extrinsic and intrinsic pathways are involved in the mechanism of apoptosis induction [65]. Tanshinone IIA has been verified to be effective for cancer suppression in osteosarcoma cells and cervical carcinoma cells, which has been reviewed in detail in the preceding part of the text. With regard to gastric cancer, it was found that tanshinone IIA reinforced antineoplastic effect in doxorubicin-resistant gastric cancer cell lines SNU-719R and SNU-610R. The inhibition of multidrug resistance-associated protein 1 was verified to increase apoptosis and induce autophagic cell death, which contributed to the potent anticancer effect of tanshinone IIA in doxorubicin-resistant gastric cancer cells [66]. Clitocine is a naturally occurring compound purified from wild mushroom and has been recently demonstrated as an apoptosis initiator in multidrug-resistant human cancer cells. A study recently found that clitocine treatment dramatically enhanced TRAIL lethality via induction of apoptosis in resistant human colon cancer cells. The disruption of the binding between Mcl-1 and Bak and mitochondrial translocation of Bax mediated by clitocine were identified as the key underlying mechanism, which in turn generated MMP. These findings indicated that clitocine was an effective adjuvant alternative in TRAIL-based cancer therapy [67]. The combination of Chinese herbal medicines with standard therapeutics has been increasingly popular for the potentiation of curative effects. Curcumin is extractable from a broad range of traditional Chinese herbs and has long been used in clinic for its capabilities of tumor inhibition. A group in the field found that compared with single treatment, combination treatment of curcumin and cetuximab dramatically induced the activation of caspase-3 and caspase-9, which are critical factors in apoptosis process [68]. Besides, another type of herb, celastrol, was studied in oral cancer. It was also studied in liver cancer, osteosarcoma, and mantle cell lymphoma as reviewed in the preceding text. Here the authors reported that exposure to celastrol led to upregulated expression of cleaved caspase-3, caspase-8, caspase-9, and PARP and downregulated expression of Bcl-2, suggesting that celastrol exerted antitumor capacity in multidrug-resistant oral cancer cells via intrinsic and extrinsic pathways [69]. Honokiol is one of the main physiologically bioactive constituents of the traditional Chinese medicine *Houpo*. A recent study showed that exposure of human U87 MG glioma cells to honokiol significantly enhanced temozolomide-induced apoptotic insults to glioma cells via an intrinsic mitochondrion-dependent mechanism, as evidenced by the enhanced activity of caspase-9 without affecting Fas and caspase-8 expression. In addition, honokiol enhanced the changes of temozolomide-induced regulation in Bax translocation, MMP, mitochondrial complex I enzyme activity, intracellular ROS level, and cytochrome c release. All these data suggested the therapeutic potential of honokiol to attenuate temozolomide-induced side effects [70].

5. Derivatives of compounds from Chinese herbal medicine that are explored for cancer treatment

Substantial native compounds are identified to possess inhibitory effects against multiple cancers. To provide therapeutic alternatives for cancer therapy and develop more efficient and specific cancer treatments, more similar compounds are being produced from naturally occurring constituents of medicinal herbs (**Table 4**).

Name of Chinese herbal medicine (active constituent)	Cancer type	Derivatives	Ref.
Artemisia annua (artemisinin)	Prostate, lung, cervical, ovarian, and breast cancers	Dihydroartemisinin	[71]
Ardisia gigantifolia (triterpenoid saponin)	Breast cancer	AG36	[72]
Pulsatilla (*Pulsatilla saponin* D)	Non-small cell lung cancer	Compound 14	[73]
Genera *Hypericum* and *Garcinia* (guttiferone)	Leukemia	Compound 2	[74]
Citrus reticulata (benzofuran)	Chondrosarcoma	BL-038	[75]
Isodon rudescens (oridonin)	Esophagus cancer	Compound 19	[76]
White birch (betulinic acid)	Prostate, gastric, and melanotic cancers	Compound 3k	[77]

Table 4.
Derivatives of compounds from Chinese herbal medicines for cancer treatment.

The natural extract artemisinin has been gaining great attention in the medical field since Chinese scientist Tu Youyou was granted Nobel prize for his discovery of artemisinin. A study lately reported that derivatives of artemisinin have great antineoplastic activities. Researchers synthesized dihydroartemisinin (DHA) and applied it in the management of tumor cell lines including PC-3, A549, HeLa, OVCAR-3, and MCF-7. The results showed that combination of DHA and doxorubicin markedly regulated the caspase cascade through the intrinsic apoptotic pathway. DHA and doxorubicin also had a significant favorable effect in vivo. This study suggested that DHA might be a potential therapeutic agent against several types of cancer [71]. AG36 is the biotransformation product of triterpenoid saponin from *Ardisia gigantifolia* Stapf. The antitumor activity and underlying molecular mechanisms of AG36 against human breast cancer cells including cell lines MCF-7, MDA-MB-231, and SK-BR-3 were investigated. Researchers found that compared with control group, the ratio of Bax/Bcl-2 and the release of cytochrome c into cytoplasm were dramatically upregulated. Western blot analysis showed that the death receptor-related proteins Fas/FasL, TNFR1, and DR5 were modulated in different breast cancer cells. This study provided a novel idea that AG36 could be used as a clinical medication against human breast cancer with regulation of extrinsic death receptor and intrinsic mitochondrial pathways [72]. A total of 17 derivatives was synthesized on the basis of molecular formula of pulsatilla saponin D. The study indicated that compound 14 induced typical cell cycle arrest and apoptosis in lung cancer A549 cells, and western blot assay suggested the involvement of both intrinsic and extrinsic apoptosis pathways in the mode of action. These data indicated that compound 14 was a potential candidate for developing new anti-lung cancer agents in the future [73]. A team recently synthesized three different types of polycyclic polyprenylated acylphloroglucinol (PPAP). Compound 2 was found to activate the intrinsic pathway by reducing the expression of anti-apoptotic protein Bcl-2 while enhancing the pro-apoptotic protein Bax. Moreover, caspase-3 and PPRP1 levels were upregulated. The present results suggested that compound 2 may merit further development as a potential antileukemia agent [74]. BL-038, the novel benzofuran derivative, has been evaluated for its antitumor activity in human chondrosarcoma cells. Chondrosarcoma is a highly malignant cartilage-forming bone tumor that is intrinsically resistant to conventional chemotherapy or radiotherapy. Recent research reported that intrinsic apoptosis response was elicited by BL-038 with the observation of release of cytochrome c, activation of caspase-9 and caspase-3, as well as the cleavage of PARP [75]. *Isodon rudescens* is well known for its antibacterial and antitumor activities and has also been regarded as a traditional green tea for centuries. Oridonin is the major bioactive ent-kaurane diterpenoid of this medicinal tea. Herein, 22 novel derivatives of oridonin were designed and synthesized. Among these compounds, compound 19 was reported to induce MOMP, which was probably involved in the intrinsic apoptotic pathway [76]. A total of 25 derivatives of betulinic acid was synthesized at C-28 position after structural modifications. The antitumor activities of these new products against human cancer cell lines including MGC-803, PC3, Bcap-37, A375, and MCF-7 were evaluated. Most of the derivatives possessed significant antiproliferative capacities. In addition, the study indicated that the apoptosis of MGC-803 cells induced by compound 3 k was mediated by mitochondrial intrinsic pathway [77]. Natural compounds have promising activities but are also quickly metabolized in the human body, leading to limited therapeutic outcomes especially in the treatment of cancer. The compound n-butylidenephthalide (BP) is isolated from *Angelica sinensis*, which has long been used as a traditional Chinese herb to treat anemia and gynecological dysfunctions. However, BP is quickly metabolized by the liver within 24 h; here an investigation prepared BP through encapsulation with a novel polycationic liposome containing polyethylenimine (PEI) and polyethylene glycol complex (LPPC) in melanoma cells. The results demonstrated that compared with BP alone treatment, BP/LPPC presented higher cytotoxicity in B16/F10 melanoma cells. BP/LPPC-treated cells showed an increase in subG1 percentage and TUNEL positive apoptotic morphology through induction of extrinsic and intrinsic apoptosis pathways [78].

6. Conclusions and perspectives

Dysregulation of programmed cell death acts as a natural barrier in survival and dissemination of cancer cells, whereas, malignant cells evolve many tricks to modify or generate some key modulators to evade programmed cell death. Apoptosis is one of the primary programmed cell death mechanisms, and extensive reports have identified intrinsic and extrinsic apoptosis pathways in cells. Its role in tumor proliferation is rather complex as different apoptotic pathways may cross talk or coexist in cancer. The decision taken by a cell to undergo apoptosis is regulated by various factors such as exogenous or endogenous damage.

For a long time, complementary medicine, in particular traditional Chinese medicine, has been extensively employed in practice due to good outcomes for patients in the treatment of either serious diseases or ailments. Recent attention has increasingly focused on the function of Chinese medicine on cancer and relevant molecular mechanisms. Chinese medicinal herbs may be of great value in the management of malignancies. This review retrospectively documented experimental data and precise effects of various herbs on tumor biology, especially the roles of apoptotic pathways in modulatory processes. Chinese herbal medicine serves to correct internal disequilibriums via the modulation of apoptosis and eventually contributes to cancer repression.

As the field of Chinese herbal medicine develops rapidly and apoptosis is not the only type of programmed cell death, which has expanded to include autophagy, and necrosis, among others, novel mechanisms of action that may be favorable in oncotherapy are still being characterized. We anticipate a major focus will be placed on other programmed cell death mechanisms as well as investigating potential functions of Chinese herbal medicines in such death pathways.

Acknowledgements

The study was financially supported by grants from the research council of the University of Hong Kong (Project Codes: 104004092, 104004460, 104004746), the Research Grants Committee (RGC) of Hong Kong, HKSAR (Project Codes: 764708, 766211, 17152116), Wong's Donation on Modern Oncology of Chinese Medicine (Project code: 200006276), Gala Family Trust (Project Code: 200007008), and Innovation Technology Fund of Hong Kong (ITF. Project code: 260900263).

Author details

Feiyu Chen, Zhangfeng Zhong, Hor Yue Tan, Ning Wang and Yibin Feng*
School of Chinese Medicine, The University of Hong Kong, Hong Kong, PR China

*Address all correspondence to: yfeng@hku.hk

References

[1] Galluzzi L et al. Cell death modalities: Classification and pathophysiological implications. Cell Death and Differentiation. 2007;**14**(7):1237-1243

[2] Galluzzi L, Kepp O, Kroemer G. Mitochondrial regulation of cell death: A phylogenetically conserved control. Microbial Cell. 2016;**3**(3):101-108

[3] Galluzzi L et al. Molecular mechanisms of cell death: Recommendations of the Nomenclature Committee on Cell Death. Cell Death and Differentiation. 2018;**25**(3):486-541

[4] Hassan M et al. Apoptosis and molecular targeting therapy in cancer. BioMed Research International. 2014;**2014**:150845

[5] Goldar S et al. Molecular mechanisms of apoptosis and roles in cancer development and treatment. Asian Pacific Journal of Cancer Prevention. 2015;**16**(6):2129-2144

[6] Pistritto G et al. Apoptosis as anticancer mechanism: Function and dysfunction of its modulators and targeted therapeutic strategies. Aging (Albany NY). 2016;**8**(4):603-619

[7] Elmore S. Apoptosis: A review of programmed cell death. Toxicologic Pathology. 2007;**35**(4):495-516

[8] Zhuang H et al. Glycine decarboxylase induces autophagy and is downregulated by miRNA-30d-5p in hepatocellular carcinoma. Cell Death & Disease. 2019;**10**(3):192

[9] Chang J et al. Molecular mechanisms of Polyphyllin I-induced apoptosis and reversal of the epithelial-mesenchymal transition in human osteosarcoma cells. Journal of Ethnopharmacology. 2015;**170**:117-127

[10] Huang ST et al. Tanshinone IIA induces intrinsic apoptosis in osteosarcoma cells both in vivo and in vitro associated with mitochondrial dysfunction. Scientific Reports. 2017;**7**:40382

[11] Pan TL et al. Proteomic analysis reveals tanshinone IIA enhances apoptosis of advanced cervix carcinoma CaSki cells through mitochondria intrinsic and endoplasmic reticulum stress pathways. Proteomics. 2013;**13**(23-24):3411-3423

[12] Chen X et al. Reactive oxygen species induced by icaritin promote DNA strand breaks and apoptosis in human cervical cancer cells. Oncology Reports. 2019;**41**(2):765-778

[13] Zhang Y et al. Flavonoids from Chinese bayberry leaves induced apoptosis and G1 cell cycle arrest via Erk pathway in ovarian cancer cells. European Journal of Medicinal Chemistry. 2018;**147**:218-226

[14] Li L et al. Phytoestrogen bakuchiol exhibits in vitro and in vivo anti-breast cancer effects by inducing S phase arrest and apoptosis. Frontiers in Pharmacology. 2016;**7**:128

[15] Kim D, Radin D, Leonardi D. Probing the molecular mechanisms governing the oncolytic activity of Paeonia suffruticosa on triple-negative breast cancer cells in vitro. Anticancer Research. 2017;**37**(9):4813-4819

[16] Zhang XL, Wang BB, Mo JS. Puerarin 6"-O-xyloside possesses significant antitumor activities on colon cancer through inducing apoptosis. Oncology Letters. 2018;**16**(5):5557-5564

[17] Zhang Y et al. Effective cytotoxic activity of OSW-1 on colon cancer by inducing apoptosis in vitro and in vivo. Oncology Reports. 2017;**37**(6):3509-3519

[18] Liu Y et al. Ginkgolic acid induces interplay between apoptosis and autophagy regulated by ROS generation in colon cancer. Biochemical and Biophysical Research Communications. 2018;**498**(1):246-253

[19] Liu M et al. In vitro assessment of *Macleaya cordata* crude extract bioactivity and anticancer properties in normal and cancerous human lung cells. Experimental and Toxicologic Pathology. 2013;**65**(6):775-787

[20] Gong X et al. Sanguinarine triggers intrinsic apoptosis to suppress colorectal cancer growth through disassociation between STRAP and MELK. BMC Cancer. 2018;**18**(1):578

[21] Cunningham KG et al. Cordycepin, a metabolic product isolated from cultures of *Cordyceps militaris* (Linn.) link. Nature. 1950;**166**(4231):949

[22] Zhang Y et al. Cordycepin induces apoptosis in human pancreatic cancer cells via the mitochondrial-mediated intrinsic pathway and suppresses tumor growth in vivo. OncoTargets and Therapy. 2018;**11**:4479-4490

[23] Liao KF et al. Anti-cancer effects of radix *Angelica sinensis* (Danggui) and N-butylidenephthalide on gastric cancer: Implications for REDD1 activation and mTOR inhibition. Cellular Physiology and Biochemistry. 2018;**48**(6):2231-2246

[24] Lu X et al. Luteolin induces apoptosis in vitro through suppressing the MAPK and PI3K signaling pathways in gastric cancer. Oncology Letters. 2017;**14**(2):1993-2000

[25] Gravitz L. Liver cancer. Nature. 2014;**516**(7529):S1

[26] Yang H et al. Celastrol, a triterpene extracted from the Chinese "Thunder of God Vine," is a potent proteasome inhibitor and suppresses human prostate cancer growth in nude mice. Cancer Research. 2006;**66**(9):4758-4765

[27] Ren B et al. Celastrol induces apoptosis in hepatocellular carcinoma cells via targeting ER-stress/UPR. Oncotarget. 2017;**8**(54):93039-93050

[28] Chang W et al. Protective effects of Celastrol on diethylnitrosamine-induced hepatocellular carcinoma in rats and its mechanisms. European Journal of Pharmacology. 2016;**784**:173-180

[29] Li H et al. Zhiheshouwu ethanol extract induces intrinsic apoptosis and reduces unsaturated fatty acids via SREBP1 pathway in hepatocellular carcinoma cells. Food and Chemical Toxicology. 2018;**119**:169-175

[30] Zhao L et al. Apoptosis induction of dehydrobruceine B on two kinds of human lung cancer cell lines through mitochondrial-dependent pathway. Phytomedicine. 2016;**23**(2):114-122

[31] Wang XD et al. Induction of apoptosis in human leukemia cells through an intrinsic pathway by cathachunine, a unique alkaloid isolated from *Catharanthus roseus*. Phytomedicine. 2016;**23**(6):641-653

[32] Liu M et al. Cytotoxicity of the compounds isolated from *Pulsatilla chinensis* saponins and apoptosis induced by 23-hydroxybetulinic acid. Pharmaceutical Biology. 2015;**53**(1):1-9

[33] Xu X et al. Coptidis rhizoma induces intrinsic apoptosis through BAX and BAK activation in human melanoma. Oncology Reports. 2017;**38**(1):538-544

[34] Wang N et al. Berberine induces autophagic cell death and mitochondrial apoptosis in liver cancer cells: The cellular mechanism. Journal of Cellular Biochemistry. 2010;**111**(6):1426-1436

[35] Choi MS et al. Berberine inhibits p53-dependent cell growth through induction of apoptosis of prostate cancer cells. International Journal of Oncology. 2009;**34**(5):1221-1230

[36] Choi MS et al. Berberine inhibits human neuroblastoma cell growth through induction of p53-dependent apoptosis. Anticancer Research. 2008;**28**(6A):3777-3784

[37] Wang N et al. Berberine-induced tumor suppressor p53 up-regulation gets involved in the regulatory network of MIR-23a in hepatocellular carcinoma. Biochimica et Biophysica Acta. 2014;**1839**(9):849-857

[38] Tan HY et al. Autophagy-induced RelB/p52 activation mediates tumour-associated macrophage repolarisation and suppression of hepatocellular carcinoma by natural compound baicalin. Cell Death & Disease. 2015;**6**:e1942

[39] Wang N et al. Inhibition of eukaryotic elongation factor-2 confers to tumor suppression by a herbal formulation Huanglian-Jiedu decoction in human hepatocellular carcinoma. Journal of Ethnopharmacology. 2015;**164**:309-318

[40] Yang PY et al. Butein induces apoptotic cell death of human cervical cancer cells. Oncology Letters. 2018;**16**(5):6615-6623

[41] Xiong Y, Wu X, Rao L. *Tetrastigma hemsleyanum* (Sanyeqing) root tuber extracts induces apoptosis in human cervical carcinoma HeLa cells. Journal of Ethnopharmacology. 2015;**165**:46-53

[42] Yang PY et al. Butein shows cytotoxic effects and induces apoptosis in human ovarian cancer cells. American Journal of Chinese Medicine. 2015;**43**(4):769-782

[43] Fu Y et al. Prodelphinidins isolated from Chinese bayberry leaves induces apoptosis via the p53-dependent signaling pathways in OVCAR-3 human ovarian cancer cells. Oncology Letters. 2017;**13**(5):3210-3218

[44] Zhu X et al. Ziyuglycoside I inhibits the proliferation of MDA-MB-231 breast carcinoma cells through inducing p53-mediated G2/M cell cycle arrest and intrinsic/extrinsic apoptosis. International Journal of Molecular Sciences. 2016;**17**(11):1903

[45] Xu Y et al. Anti-proliferative effects of paeonol on human prostate cancer cell lines DU145 and PC-3. Journal of Physiology and Biochemistry. 2017;**73**(2):157-165

[46] Cao C et al. Ginkgo biloba exocarp extracts induces apoptosis in Lewis lung cancer cells involving MAPK signaling pathways. Journal of Ethnopharmacology. 2017;**198**:379-388

[47] Chen CC et al. The anti-cancer effects and mechanisms of *Scutellaria barbata* D. Don on CL1-5 lung cancer cells. Oncotarget. 2017;**8**(65):109340-109357

[48] Yu CY et al. Ovatodiolide isolated from *Anisomeles indica* induces cell cycle G2/M arrest and apoptosis via a ROS-dependent ATM/ATR signaling pathways. European Journal of Pharmacology. 2018;**819**:16-29

[49] Yang ST et al. Bisdemethoxycurcumin-induced S phase arrest through the inhibition of cyclin A and E and induction of apoptosis via endoplasmic reticulum stress and mitochondria-dependent pathways in human lung cancer NCI H460 cells. Environmental Toxicology. 2016;**31**(12):1899-1908

[50] Hu A et al. Curcumin as therapeutics for the treatment of head and neck squamous cell carcinoma by activating SIRT1. Scientific Reports. 2015;**5**:13429

[51] Zhu GH et al. Curcumin induces apoptosis and suppresses invasion through MAPK and MMP signaling in human monocytic leukemia SHI-1 cells. Pharmaceutical Biology. 2016;**54**(8):1303-1311

[52] Li HY et al. Celastrol induces apoptosis and autophagy via the ROS/JNK signaling pathway in human osteosarcoma cells: An in vitro and in vivo study. Cell Death & Disease. 2015;**6**:e1604

[53] Xie Y et al. Dihydrocelastrol exerts potent antitumor activity in mantle cell lymphoma cells via dual inhibition of mTORC1 and mTORC2. International Journal of Oncology. 2018;**53**(2):823-834

[54] Oh H et al. Licochalcone B induces apoptosis of human oral squamous cell carcinoma through the extrinsic- and intrinsic-signaling pathways. International Journal of Oncology. 2016;**48**(4):1749-1757

[55] Kang TH et al. Natural compound licochalcone B induced extrinsic and intrinsic apoptosis in human skin melanoma (A375) and squamous cell carcinoma (A431) cells. Phytotherapy Research. 2017;**31**(12):1858-1867

[56] Zhang Y et al. Downregulation of STAT3 and activation of MAPK are involved in the induction of apoptosis by HNK in glioblastoma cell line U87. Oncology Reports. 2014;**32**(5):2038-2046

[57] Li C et al. Allicin induces apoptosis through activation of both intrinsic and extrinsic pathways in glioma cells. Molecular Medicine Reports. 2018;**17**(4):5976-5981

[58] Chen YC et al. Hispolon from *Phellinus linteus* induces G0/G1 cell cycle arrest and apoptosis in NB4 human leukaemia cells. American Journal of Chinese Medicine. 2013;**41**(6):1439-1457

[59] Taechakulwanijya N et al. Apoptosis-inducing effects of jujube (Zao) seed extracts on human Jurkat leukemia T cells. Chinese Medicine. 2016;**11**:15

[60] Zhao X et al. Aqueous extracts of Cordyceps kyushuensis kob induce apoptosis to exert anticancer activity. BioMed Research International. 2018;**2018**:8412098

[61] Chang HP et al. Pterostilbene modulates the suppression of multidrug resistance protein 1 and triggers autophagic and apoptotic mechanisms in cisplatin-resistant human oral cancer CAR cells via AKT signaling. International Journal of Oncology. 2018;**52**(5):1504-1514

[62] Zhong ZF et al. Furanodiene induces extrinsic and intrinsic apoptosis in doxorubicin-resistant MCF-7 breast cancer cells via NF-kappaB-independent mechanism. Frontiers in Pharmacology. 2017;**8**:648

[63] You C et al. Trichosanthin enhances sensitivity of non-small cell lung cancer (NSCLC) TRAIL-resistance cells. International Journal of Biological Sciences. 2018;**14**(2):217-227

[64] Tao Y et al. Huaier augmented the chemotherapeutic sensitivity of oxaliplatin via downregulation of YAP in hepatocellular carcinoma. Journal of Cancer. 2018;**9**(21):3962-3970

[65] Fu Z et al. Euphorbia lunulata extract acts on multidrug resistant gastric cancer cells to inhibit cell proliferation, migration and invasion, arrest cell cycle progression, and induce apoptosis. Journal of Ethnopharmacology. 2018;**212**:8-17

[66] Xu Z et al. Potentiation of the anticancer effect of doxorubicinin drug-resistant gastric cancer cells by tanshinone IIA. Phytomedicine. 2018;**51**:58-67

[67] Sun JG et al. Clitocine potentiates TRAIL-mediated apoptosis in human colon cancer cells by promoting Mcl-1 degradation. Apoptosis. 2016;**21**(10):1144-1157

[68] Son DJ et al. Synergistic inhibitory effects of cetuximab and curcumin on human cisplatin-resistant oral cancer CAR cells through intrinsic apoptotic process. Oncology Letters. 2018;**16**(5):6323-6330

[69] Lin FZ et al. Celastrol induces vincristine multidrug resistance oral cancer cell apoptosis by targeting JNK1/2 signaling pathway. Phytomedicine. 2019;**54**:1-8

[70] Chio CC et al. Honokiol enhances temozolomide-induced apoptotic insults to malignant glioma cells via an intrinsic mitochondrion-dependent pathway. Phytomedicine. 2018;**49**:41-51

[71] Tai X et al. In vitro and in vivo inhibition of tumor cell viability by combined dihydroartemisinin and doxorubicin treatment, and the underlying mechanism. Oncology Letters. 2016;**12**(5):3701-3706

[72] Mu LH et al. AG36 inhibits human breast cancer cells proliferation by promotion of apoptosis in vitro and in vivo. Frontiers in Pharmacology. 2017;**8**:15

[73] Chen Z et al. Cytotoxicity, hemolytic toxicity, and mechanism of action of Pulsatilla Saponin D and its synthetic derivatives. Journal of Natural Products. 2018;**81**(3):465-474

[74] Lin X et al. Synthesis of novel guttiferone E and xanthochymol derivatives with cytotoxicities by inducing cell apoptosis and arresting the cell cycle phase. European Journal of Medicinal Chemistry. 2019;**162**:765-780

[75] Liu JF et al. BL-038, a Benzofuran derivative, induces cell apoptosis in human chondrosarcoma cells through reactive oxygen species/mitochondrial dysfunction and the caspases dependent pathway. International Journal of Molecular Sciences. 2016;**17**(9):1491

[76] Li D et al. Antitumor and antibacterial derivatives of oridonin: A main composition of Dong-Ling-Cao. Molecules. 2016;**21**(5):575

[77] Chakraborty B et al. Synthesis and biological evaluation of a novel betulinic acid derivative as an inducer of apoptosis in human colon carcinoma cells (HT-29). European Journal of Medicinal Chemistry. 2015;**102**:93-105

[78] Gao HW et al. Antitumor effect of n-butylidenephthalide encapsulated on B16/F10 melanoma cells in vitro with a polycationic liposome containing PEI and polyethylene glycol complex. Molecules. 2018;**23**(12):3224

Chapter 2

Programmed Cell Death Deregulation in BCR-ABL1-Negative Myeloproliferative Neoplasms

Carmen C. Diaconu, Petruta Gurban, Cristina Mambet, Mihaela Chivu-Economescu, Laura G. Necula, Lilia Matei, Denisa Dragu, Saviana Nedeianu, Ana I. Neagu, Aurelia Tatic, Diana Cristodor and Coralia Bleotu

Abstract

BCR-ABL1-negative myeloproliferative neoplasms are classically represented by primary myelofibrosis, polycythemia vera, and essential thrombocythemia. These entities are stem cell-derived clonal disorders characterized by hematopoietic progenitor autonomy or hypersensitivity to cytokines, most of them presenting mutations in Janus kinase 2 (*JAK2*), calreticulin (*CALR*), or myeloproliferative leukemia virus oncogene (*MPL*). Deregulation of pro- and antiapoptotic genes is also claimed as an important mechanism involved in cell resistance to cell death and accumulation of myeloid cells in myeloproliferative neoplasms. Apoptosis, as one of the best-characterized types of programmed cell death, has a clear role in hematopoiesis control. However, the exact pathways affected in BCR-ABL1-negative myeloproliferative neoplasms have not yet been fully clarified. This chapter will explore the modifications affecting programmed cell death pathways involved in myeloid proliferation and how these alterations might be exploited in single or combined targeted therapeutic strategies.

Keywords: apoptosis, programmed cell death, cancer stem cells, hematological disorders, cell death mechanisms, molecular interactions, cytokine signaling, cell-cycle inhibitors

1. Introduction

Hematopoiesis is a highly controlled process that ensures the differentiation of the hematopoietic stem cells (HSCs) into lymphoid and myeloid common progenitors and further into all lineages of blood cells [1].

Programmed cell death (PCD) is one of the fundamental mechanisms of an organism's life cycle that controls every system, including hematopoietic system, based on a precisely tuned signaling network. Apoptosis, the most important type of PCD, maybe because it is the most analyzed type of death to date, is well described in hematopoietic differentiation [2]. Its deregulation in pathological circumstances is potentially deleterious and may influence the fate of the entire organism. Although different other types of PCD were described, apoptosis remains one of the most important processes involved in differentiation and cell survival regulation, while mechanisms as autophagy and necroptosis look like "backup" mechanisms that share some "key players" and diverged from apoptosis at a certain point, to assure the elimination of the malfunctioning system in case of "internal" defect (mutations) or pathogens that inhibit the components of the apoptotic network [3, 4].

Primary myelofibrosis (PMF), polycythemia vera (PV), and essential thrombocythemia (ET) are classic BCR-ABL1-negative myeloproliferative neoplasms (MPNs) that are stem cell-derived clonal disorders characterized by hematopoietic progenitor autonomy or hypersensitivity to cytokines, driven by acquired somatic mutations in critical pathways, resulting in pathological expansion of the myeloid lineages. In their natural course, MPNs could be exacerbated and transformed into secondary acute myeloid leukemia (sAML) associated with treatment resistance and poor clinical outcome [5].

This chapter will explore the most important modifications affecting programmed cell death pathways involved in myeloid proliferation, and how these alterations might be exploited in single or combined targeted therapeutic strategies in a classic BCR-ABL1-negative MPN.

2. Intrinsic and/or extrinsic apoptotic pathways involved in MPN disease entities

Overall, the hematopoietic homeostasis requires a precise balance between blood cell formation and maintenance of an adequate number of mature cells. Although apoptosis is necessary to prevent the excessive accumulation of cells, the hematopoietic progenitors need to be protected and preserved. A disruption of the homeostatic balance in the hematopoietic system is relevant for many hematological disorders, an increased cell death being involved in the etiology of immune deficiencies and anemia, while an inappropriate resistance to apoptosis might lead to hematological malignancies [6], such as MPNs.

As a particular form of PCD, apoptosis is activated via two convergent pathways: the intrinsic and the extrinsic [7]. The intrinsic signaling pathway is triggered at mitochondrial level in response to various stimuli such as genotoxic agents or growth factor deprivation, and it is mainly regulated by the members of BCL-2 protein family that contain one or more BCL2 homology (BH) domains [8]. These proteins are structurally and functionally classified into three groups. The first group includes the critical effectors of the intrinsic pathway, namely BCL-2 antagonist killer 1 (BAK) and BCL-2-associated X protein (BAX). The second group is represented by the prosurvival BCL-2 proteins (BCL-2, BCL-xL, MCL-1, BCL-W, and A1) that hinder BAK and BAX activation, while the third group comprises several structurally different proteins, known as "BH3-only" proteins (BIM, BID, BAD, BIK, PUMA, and NOXA), which share solely a sequence called BH3-domain [8, 9]. Cellular stress signals are sensed by the "BH3-only" proteins that directly activate BAK and BAX or neutralize the prosurvival proteins. Once activated, through conformational changes, BAK and BAX induce the permeabilization of the mitochondrial membrane with subsequent release of apoptogenic factors, such as cytochrome c and second mitochondrial activator of caspases/direct IAP binding protein with low pI (SMAC/DIABLO). Cytochrome c binds to the apoptotic protease activating factor-1 (APAF-1) and forms the apoptosome, a heptameric complex that activates the initiator caspase-9, followed by activation of the effector caspase-3, caspase-6, and caspase-7 that trigger final events of apoptosis [10, 11].

In the extrinsic apoptotic pathway, caspase activation is elicited at the level of "death receptors" (DR), transmembrane proteins of the tumor necrosis factor (TNF) receptor superfamily typically represented by FAS (CD95), TNF receptors, and TNF-related, apoptosis-inducing ligand (TRAIL) receptors. Through interaction with their corresponding ligands—FASL, TNF-α, and TRAIL, respectively—DR become activated, leading to the recruitment of a death adaptor protein, such as FAS-associated death domain (FADD) or TNFR-1-associated death domain (TRADD). Death adaptors generate a death-inducing signaling complex (DISC), in which procaspase-8 is recruited and activated, the death signal being subsequently transduced to the effector caspases [2, 10].

A very early apoptosis event is the global and rapid mRNA degradation by a mechanism that is not yet completely characterized [12].

Various factors associated with intrinsic and extrinsic apoptotic pathways have been involved in the control of adult hematopoiesis under physiological as well as pathological conditions [2]. In this respect, BCL-2 protein family members play different roles across individual hematopoietic lineages during differentiation and maturation. At the level of HSCs and early myeloid progenitors, MCL-1 is an essential prosurvival factor, being upregulated by stem cell factor and interleukin-3 via JAK/STAT (Janus-activated kinase/signal transducers and activators of transcription) and AKT signaling pathways [13, 14]. During erythropoiesis, erythropoietin (EPO) ensures erythroid progenitor survival, proliferation, and differentiation by acting on its cognate receptor (EPO-R) and inducing JAK2-STAT5 activation that leads to upregulation of BCL-xL [10]. The development, maturation, and survival of megakaryocytes (MKC) is strictly dependent on the presence of both BCL-xL and MCL-1 proteins that are induced by thrombopoietin (TPO) signaling and restrain intrinsic apoptosis, while platelet life span seems to be dictated only by BCL-xL levels [7, 15, 16]. Similarly, MCL-1 is essential for granulocyte progenitor survival and differentiation [16]. On the other hand, the receptor/ligand interactions of the TNF family represent physiological mechanisms that exert a negative regulation in the terminal stages of the hematopoietic differentiation, controlling in this way the size of the expanding hematopoietic clones and maintaining heterogeneity in response to various demands [17].

PV is characterized by erythrocytosis accompanied by a suppressed endogenous EPO production. It often associates thrombocytosis and/or leukocytosis with panmyelosis at bone marrow examination. The pattern of driver mutations is strikingly dominated by *JAK2* V627F that is present in more than 95% of patients, the rest being represented by *JAK2* exon 12 mutations [18].

A study that analyzed gene expression profile of granulocytes isolated from PV patients showed an upregulation of protease inhibitors with affinity for proteases inducing apoptosis in neutrophils (e.g., cystatin F and secretory leukocyte protease inhibitor), as well as of several antiapoptotic and survival factors (e.g., p38 MAPK), compared to granulocytes obtained from healthy subjects [19]. Also, unlike the granulocytes of ET patients or normal controls, the granulocytes of PV patients were found to express an increased amount of heat shock protein 70 (HSP70), which counteracts apoptosis at different levels by preventing BAX translocation to mitochondria, inhibiting APAF-1 and procaspase-9 recruitment to apoptosome, and reducing caspase activation. As shown in primary cell cultures, an HSP-70 inhibitor was able to induce apoptosis in the erythroid lineage [20].

Concerning the extrinsic apoptosis pathway, it was found that erythroblasts isolated from PV patients carrying *JAK2* V617F mutation exhibited an increased resistance to death receptor-induced apoptosis being able to generate elevated red blood cell counts in the presence of CD95 and TRAIL receptor stimulation. In addition, the *JAK2* mutation was correlated in PV erythroblasts with an overexpression of c-FLIPshort, a potent cellular inhibitor of extrinsic apoptosis [21]. Also, Tognon et al. reported a dysregulated expression of genes related to extrinsic apoptosis (*FAS*, *FASL*, *FAIM*, *C-FLIP*, *TRAILRl/DR4*, and *TRAILR2/DR5*) in bone marrow CD34+ cells and peripheral blood leukocytes obtained from patients with different MPN phenotypes, including PV [22].

ET is defined by thrombocytosis associated with normocellular bone marrow and hyperplasia of enlarged MKC. The molecular profile of ET consists of *JAK2* V617F mutation (in 60–65% of patients), *CALR* exon 9 indels (in 20–25% of patients), *MPL* exon 10 mutations (in about 4–5% of patients), and very rare non-canonical *MPL* mutations (in less than 1% of patients). About 10% of ET patients lack these mutations being considered triple-negative cases [18].

Before the discovery of *JAK2* V617F, in order to gain insight into the molecular mechanisms of ET megakaryopoiesis, Tenedini et al. have employed microarray technology to study the gene expression profiles of bone marrow CD34-derived MKC from ET and healthy individuals. They found in ET a downregulation of the proapoptotic genes *BAX*, *BNIP3*, and *BNIP3L*, as well as of the genes encoding for components of the mitochondrial permeability transition pore complex, along with the upregulation of the antiapoptotic and survival genes *IGF1R*, *CFLAR (C-FLIP)*, and *SDF1*. Also, ET MKC exhibited in cell cultures an increased resistance to apoptosis, relative to their normal counterparts [23].

In a study that aimed to characterize the immunophenotypic apoptotic profiles of MKC on bone marrow biopsy samples obtained from MPN patients, it was observed that ET MKC displayed an antiapoptotic pattern, characterized by an overexpression of BCL-xL and a lower expression of BAX, compared to those of PMF patients [24]. Furthermore, Treliński et al. confirmed by flow cytometry the antiapoptotic profile of ET MKC and bone marrow mononuclear cells (BMMC). As opposed to controls, previously untreated ET patients presented significantly lower percentages of apoptotic MKC and BMMCs, when assessed for the number of annexin-V+ and caspase-3+ positive cells. These findings were accompanied by markedly lower BAX levels and BAX/BCL-2 ratios, especially in *JAK2* V617F-negative cases [25].

Compared to PV and ET, PMF is a more heterogeneous disease, being characterized by clonal myeloproliferation, abnormal cytokine expression, early bone marrow fibrosis, anemia, splenomegaly, extramedullary hematopoiesis, constitutional symptoms, and a lower overall survival rate. On the other hand, during the natural course of the disease PV and ET patients might suffer a conversion into secondary myelofibrosis (MF) that resembles PMF [26, 27]. PMF shares with ET a similar profile of mutations in *JAK2*, *CALR*, and *MPL* [18].

Initially, it was suggested that bone marrow MKC in PMF might undergo an increased apoptosis that could be responsible for the release of fibrogenic cytokines [28]. However, further studies have demonstrated that PMF MKC displayed a high proliferative capacity and resistance to apoptosis, explained by the overexpression of BCL-xL [29]. Also, the gene expression analysis of laser-microdissected MKC from PMF patients indicated a tendency toward an overall downregulation of apoptosis-associated genes, especially of *BNIP3* [30].

Chronic inflammation sustained by the continuous release of proinflammatory cytokines and chemokines and subsequent bone marrow microenvironment alterations are considered key factors in PMF pathogenesis. The abnormal production of cytokines that occurs both in malignant and nonmalignant cells was related to an increased JAK2-STAT3 activation and was found responsible for the inhibition of apoptosis and increased myeloproliferation, creating an environment that favors

MPN clone maintenance and expansion [31, 32]. Recently, it was shown that MF cells downregulated the expression of X-linked inhibitor of apoptosis (XIAP) and mitogen-activated protein kinase 8 (MAPK 8), a necessary component of TNF-mediated apoptosis, via a TNF/TNFR2-dependent autocrine loop. This was considered a mechanism to escape an apoptotic response and to increase NF-κB signaling involved in inflammatory cytokine expression [33].

Overall, these data show the importance of the participation of both intrinsic and extrinsic apoptosis pathways in the pathogenesis of MPNs.

3. Key PCD players in BCR-ABL1-negative MPN entities

Modifications occurred in the regulation of apoptosis, especially in expression of pro- and antiapoptotic genes, have great contribution to the myeloaccumulation in MPNs. Concerning the involvement of other types of PCD in myeloproliferations, few data are available. Some key players are involved in apoptosis regulation and also in autophagy or necroptosis. More often, it is a continuous process from apoptosis, autophagy to necroptosis. Increased death signals and stress levels can switch cell death types in the attempt of eliminating the malfunctioning cells [34, 35].

BCL-2 family of proteins is a very important regulator of apoptosis and, at the same time, is also a negative regulator of BECN1/Beclin-1, a key regulator of autophagy [36, 37]. Autophagy was shown to be a major contributor to chemotherapy resistance in AML [38].

BCL-xL promotes cell survival, such as survival of erythroid cells and platelets, and regulates their lifespan at a steady state. Inhibition of BCL-xL induces profound thrombocytopenia by triggered BAK/BAX-mediated mitochondrial damage, caspase activation, and premature death of MKC [39, 40]. In MPNs, a concerted effect resulted from antiapoptotic BCL-xL over-expression and proapoptotic BNIP-3 downregulation was clearly documented [41].

Bcl-2-associated death promoter (BAD) inhibits antiapoptotic proteins BCL-2 and BCL-xL and is involved in initiating the apoptosis process. In unphosphorylated form, BAD forms heterodimers with BCL-2 and BCL-xL, inhibiting their antiapoptotic functions, and facilitates BAX/BAK activation in response to apoptotic stimuli [42, 43], promoting apoptosis. After activation by phosphorylation, BAD forms a heterodimer with 14-3-3 proteins, releasing BCL-2 that is free to block apoptosis. BAD is a substrate of various kinases, such as AKT, protein kinase A (PKA), and c-Jun NH2-terminal kinase (JNK).

Gene expression studies on CD34+ cells and peripheral leukocytes isolated from ET and PMF patients indicated that mRNA levels of *BAX*, *BAD,* and *BIK* were lower in *JAK2* V617F-positive cases than in negative ones and, additionally, displayed a negative correlation with the *JAK2* V617F mutational load. Also, *A1, MCL1, BCLW,* and *BCL-XL* genes have an increased expression in ET and PMF patients compared to controls. As such, Tognon et al. hypothesized that deregulated expression of apoptosis-related genes is linked to myeloaccumulation and pathogenesis of these two disorders [44].

Studies focused on the apoptosis deregulation in PV identified an increased expression of *A1* and *MCL-1* and a reduced expression of proapoptotic *BAD* and *BAX* genes in PV CD34+ cells compared with controls. *A1* expression was also increased, whereas *BAD* and *BAX* mRNA levels were decreased in the leukocytes of PV patients versus healthy subjects [45]. Rubert et al. evaluated the roles of proapoptotic BIM and antiapoptotic MCL-1 in regulating *JAK2* V617F-positive cell survival. JAK2 inhibition modified Bim-EL Ser69 phosphorylation, as well as decreased MCL-1 level, inducing apoptosis. On the other side, MCL-1 depletion compromised cells' viability and sensitized *JAK2* V617F-positive cells to JAK2 inhibition [46]. Also, mutant JAK2 inhibits the BCL-xL deamidation pathway and the apoptotic response to DNA damage in primary cells from patients PV [47]. *JAK2* V617F-positive ET patients presented markedly higher activation of caspase-3, as well as higher BAX expression than *JAK2* V617F-negative ones [25]. Recent data pointed out that in ET, the MKC exhibit a more proliferative profile, while in MF, they display, in a larger proportion, a defective proapoptotic mechanism [24, 48].

Survivin is one of the inhibitors of apoptosis proteins (IAPs) that regulate cell death through mitochondrial route by restricting the IAP-inhibitor DIABLO protein and preventing it from activating caspase-9. A greater proportion of myeloproliferative MKC express survivin compared to its reciprocal inhibitor, DIABLO. Survivin seems to be the key mediator of the MKC survival signature in the MPNs and might be a potential therapeutic target [41]. Recently, new evidence suggested that survivin may be involved in the evasion of cell death by manipulation of autophagy [49].

BNIP-3 (BCL2/adenovirus E1B 19 kDa protein-interacting protein 3), a pro-apoptotic mitochondrial protein belonging to the BCL-2 family, is activated under hypoxic conditions with hypoxia-inducible factor (HIF-1α) in normal and cancer tissues. BNIP-3 is involved in the induction of hypoxic necrosis in tumors because it activates caspase-independent necrosis-like cell death by opening the mitochondrial permeability transition pore. In MPNs, BNIP-3 expression is reportedly low and this might indicate that the increased bone marrow cellularity is not only because of proliferative signaling but also due to decreased apoptosis [50, 51].

CALR is a multifunctional endoplasmic reticulum (ER) chaperone involved in the quality control of N-glycosylated proteins, calcium storage, and immune responses [27, 52]. In relation to apoptosis, CALR is implicated in the specific activation of caspase-8, BAX, and BAK, and also in the BCL-2 cleavage [53].

Caspase-8, a key factor in the extrinsic pathway, together with caspase-9, a key factor in the intrinsic pathway, is implicated in regulating MKC turnover [41]. *CALR* mutations particularly affect the MKC lineage as indicated by the higher mean number of endogenous MKC colonies in *CALR*-mutant MPNs than those found in *JAK2* V617F-positive and triple-negative cases [54]. Immunohistochemical studies proved that *CALR*-mutated MKC displayed a dysregulated apoptosis with significant reductions in proapoptotic BNIP-3 that could explain the higher platelet number observed in *CALR*-positive MPNs, in contrast to other molecular subtypes [48]. Caspase-8 just like caspase-9 regulates MKC turnover in the MPNs. Overexpression of caspase-8 induces *TP53* gene transcription to produce p53, which stimulates apoptotic cascade [41]. Caspase-8 uses proteolytic and nonproteolytic functions to change cell behavior. Activated caspase-8 triggers caspase-3 activation that commits cell to death [55]. (How does this relate to MPN? Is it upregulated or downregulated?).

Immunohistochemistry studies showed that the percentage of MKC positive for caspase-8 is higher in MPNs in comparison with controls, suggesting that MKC in MPN tend to counteract the survival advantages acquired through inhibition of the intrinsic apoptotic pathway by activating the caspase-8-mediated extrinsic apoptotic cascade[41].

Caspase-9 is an inducible proapoptotic molecule, which acts relatively late in apoptosis signaling becoming less susceptible to inhibition by apoptosis inhibitors [56]. Caspase-9 is an apoptotic initiator caspase in MKC and platelets being necessary for their efficient death, and it is not required for platelet generation and function, as it was previously thought. Thus, caspase-9 loss is associated with an increased MKC proliferative capacity. In MPNs, especially in the *CALR*-mutated molecular subtype, caspase-9 dysfunction could play a role in the enhanced thrombocytosis [40, 41].

SMAC/DIABLO controls apoptosis by negatively regulating IAPs and by activating caspases. Recently, it was shown that silencing of SMAC/DIABLO caused decreased levels of phospholipids, suggesting that besides proapoptotic functions, SMAC/DIABLO have nonapoptotic lipid synthesis-related function essential for cancer growth and development. Therefore, it was assumed that SMAC/DIABLO could be a promising therapeutic target in cancer [57]. On the other hand, SMAC/DIABLO downregulation was found to be associated with progressive disease and poor survival rate in hematologic malignancies, and DIABLO/SMAC mimetics were proposed as a potential adjunct therapy to enhance DIABLO levels in MPN MKC [41]. More studies are necessary to establish the proper therapeutic options in the light of the new role of SMAC/DIABLO in the phospholipid synthesis.

The tumor suppressor gene *TP53* plays many roles in apoptotic landscape by suppressing or activating a large number of checkpoint and apoptotic genes [58]. Alterations in *TP53* have not been linked to MKC hyperplasia although mutations targeting *TP53* do occur during leukemic transformation of MPNs. Mutated p53 could inhibit the wild-type p53 function or gain new oncogenic functions through protein-protein interactions [59]. As a result, poor prognosis in hematologic malignancies is correlated with mutations in *TP53* due to mutant's stability [58]. Elevated levels of MKC p53-positive are present more in PMF than in PV and ET [41]. According to Malherbe et al., *CALR*-mutated cases compared to *JAK2* V617F-positive cases present more p53, but not caspase-8 positivity. It seems that CALR lesions disrupt alternative apoptotic effectors and affected MKC attempt, a remedial prodeath response dominated by overexpression of p53. Also, *CALR*-mutated MKC display a minor caspase-9 upregulation that is unlikely to induce apoptosis due to concurrent survivin overexpression, but rather promote thrombocytosis [41].

Cell surface death receptor-ligand interaction, such as FASL binding FAS, TRAIL binding death receptor 5 (DR5) or TNFα binding TNFR1, executes extrinsic pathway apoptosis.

The two major necroptotic death effector complexes, the necrosome and ripoptosome, are induced by TNFR1 and toll like receptor 3 (TLR3) signaling, respectively [60]. IFN-R activation, primarily by type-I IFN, is believed to involve a caspase- and FADD-independent, receptor-interacting protein kinase (RIP) 3-dependent mode of cell death via the formation of the necrosome. Following IFN-R activation, JAK/STAT signaling and the activity of RNA-responsive protein kinase (PKR), upstream of RIP1/RIP3 necrosome formation is essential. TNFα binding to TNFR1 causes recruitment of TRADD and RIP1 via their death domains resulting in the prosurvival complex I, which is stabilized by TNFα-bound TNFR2-TRAF2. Internalization of the TNFR1-TRADD-RIP1 complex is required for recruitment of caspase-8 and FADD, necessary for apoptosis. This is therefore a major cell death checkpoint as the absence of NF-κB activation and prosurvival signaling results in proapoptotic complexes or, alternatively, the pronecroptotic complex known as the necrosome. The ripoptosome consists of FADD, cFLIP and caspase-8 and allows necroptosis to prevail if active cleavage of RIP1 by caspase-8 is prevented by cFLIPL. In MPN, it was shown that blocking TNFR2 but not TNFR1 selectively inhibits MPN cells over normal ones and the process involves XIAP, cIAP, and MAPK8 as key mediators of these differential responses to TNF. These data support the potential therapeutic use of cIAP inhibitors and selective TNFR2 inhibitors in the treatment of MF [33].

The TLR3-induced pathway converges with the TNFR1-induced pathway at the necrosome. The execution phase of necroptosis starts with interaction between RIP1 and RIP3. Following stabilization of the RIP1-RIP3 complex, mixed-lineage kinase domain-like protein (MLKL) is recruited to form a functional necrosome. MLKL activated upon phosphorylation by RIP3 results in the translocation of the

MLKL necrosome to the plasma membrane, necroptotic membrane disruption, and release of liposomes containing phosphatidylinositol phosphates (PIPs). This permeabilization, combined with MLKL-mediated calcium or sodium influx ion-pore dysregulation, characterizes the model proposed for necroptosis execution [60]. Human cancers, including MPN and their exacerbated form, sAML, are known for eluding apoptosis; therefore, therapeutic induction of necroptosis may represent a better strategy for an efficient treatment. A series of compounds have been shown to trigger necroptosis, particularly inhibitors of RIP1, RIP3 or MLKL, in leukemia cells; however, a deeper understanding of the signaling network that regulates this type of PCD is still necessary [35].

4. Major signaling pathways involved in apoptotic failure in molecular subgroups of MPN

The constitutive activation of JAK-STAT pathway is a common feature of MPNs irrespective of driving mutation, being observed even in so-called "triple-negative MPNs" that lack known *JAK2*, *MPL,* and *CALR* mutations [41, 48, 61]. In contrast to the transient activation of the JAK-STAT signaling that occurs in the physiological conditions, MPNs are characterized by a hyperactive JAK2 signaling through dimeric myeloid cytokine receptors (EPO-R, TPO-R, and G-CSFR) even in the absence of the ligand that promotes myeloproliferation and resistance to apoptosis (**Figure 1**) via the JAK-STAT, PI3K (phosphatidylinositol 3-kinase)-AKT signaling pathways, and ERK/MAPK pathways [62, 63]. While the most prevalent MPN driver mutation, *JAK2* V617F, induces the activation of all three myeloid cytokine receptors, *MPL* and *CALR* mutations activate only TPO-R. This provides

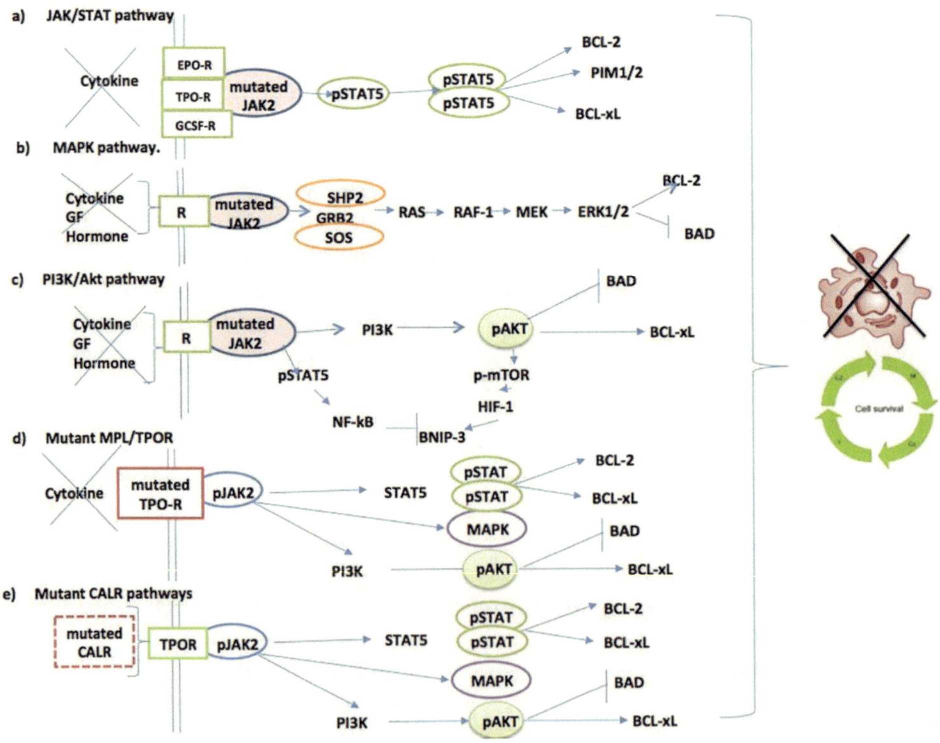

Figure 1.
Major signaling pathways involved in apoptotic failure in molecular subgroups of MPN.

an explanation for the association of *JAK2* V627F mutation with all classical MPN phenotypes (PV, ET, and PMF) and also for the preferential occurrence of *MPL* and *CALR* mutations in ET and PMF.

In addition, loss-of-function or neomorph mutations in genes that are involved in epigenetic regulation, splicing, and signaling can act as disease modifiers by cooperating with MPN driver mutations [52].

The JAK/STAT is the major pathway (**Figure 1a**) involved in MPN pathology [64–66]. *JAK2* V617F mutation promotes constitutive activation of JAK-STAT signaling, erythrocytosis and MKC proliferation, extensive cellular hyperplasia, and abrogated apoptosis [67, 68]. In response to *JAK2* V617F mutation, extensive proliferation conduces to accumulation of irreparable DNA damage. As a consequence, internal apoptotic cascade is triggered with higher BNIP-3 positivity that stimulates MKC apoptosis [69, 70]. This is counteracted by the antiapoptotic effects conferred by excess BCL-XL expression induced by phosphorylated STAT5 (pSTAT5) and pSTAT3. BCL-xL expression is essential to maintain megakaryoblast lineage survival and platelet production, preventing lethal hemorrhage [15].

The MAPK/ERK signaling pathway activation (**Figure 1b**) is required in MKC differentiation, with TPO as signal for induced maturation via MPL receptor [71]. *JAK2* V617F can activate MAPK signaling pathway via receptor tyrosine kinase-Grb2-SOS, continuing with RAS GTPase and RAF-1, which activates MEK, followed by ERK activation [72]. Phosphorylated extracellular signal-regulated kinase (ERK) activates BCL-2 and BAD, both of which have the effect of inhibiting apoptosis.

An increased activation of RAS/RAF/ERK pathway was showed in MPN patients, especially in erythroid precursor cells and MKC (**Figure 1b**). It was shown that ERK is constitutively activated by the *JAK2* V617F mutation. Laubach JP et al. demonstrated an increased activation of RAS/ERK pathway in PV, associated with a dysregulated erythropoiesis and apoptosis resistance of erythroid precursor cells [73]. ET, PV, and PMF patients with *JAK2* V617F mutation demonstrated an increase of ERK phosphorylation level in MKC. Phosphorylated ERK activates BAD, which release apoptosis inhibitor BCL-2, resulting in an overall inhibition of apoptosis that may be the cause of MKC hyperplasia and bone marrow hypercellularity [50].

The PI3K/AKT signaling pathway (**Figure 1c**) may be activated by the *JAK2* V617F mutation or pSTAT5. PI3K/AKT pathway is involved in several cellular processes including cell proliferation and differentiation, protein synthesis, and apoptosis. AKT pathway is known to be active in AML [74, 75]. Dai C et al. showed that increased erythroid progenitor proliferation from PV is associated with increased phosphorylation of AKT [76]. Khan I et al. confirmed that PI3K/AKT signaling pathway was activated in MPN by the *JAK2* V617F and *MPL* W515 L mutations [77]. Moreover, Koopmans et al. demonstrated an increase of pAKT level in the cytoplasm and nucleus of immature myeloid cells and in MKC of MPN patients. Immunohistochemical staining showed that pAKT expression was significantly higher in MKC of ET compared to PV and PMF. Recently, the AKT activation was demonstrated to be a feature of *CALR*-mutant myeloproliferative neoplasms [78]. The higher platelet counts reported in MPN with *CALR* mutations may be due to greater dysregulation of MKC apoptosis [48]. Consequently, AKT was considered a potential target in MPN therapy. Several studies have shown that targeting AKT with specific inhibitors reduced cell growth *in vitro* [77] and induced prolonged survival of the immunodeficient mice injected with *JAK2* V617F-mutated cells, along with reducing spleen size [79].

The activation of AKT upregulates BCL-xL and inactivates BAD, suppressing apoptosis and promoting cell survival. This was observed in MPN patients, where the activation of pAKT was higher in MKC and associated with the inhibition of MKC apoptosis [50]. pAKT is also known to induce activation of BNIP-3 and caspase-9 through mammalian target of rapamycin (mTOR)—a serine/threonine kinase that is an effector protein of AKT—via activation of HIF-1. Data related to BNIP-3 expression are conflicting; some groups reported a reduced BNIP-3 expression [30], whereas others have shown its upregulation in MPNs [50]. In the study of Koopmans et al., the immunohistochemical expression of BNIP-3, with proapoptotic function, was lower in total bone marrow cells of ET, PV, and PMF patients, compared with the control group. This suggests that a decreased apoptosis might also contribute to the increased bone marrow cellularity observed in MPNs. However, in contrast to total bone marrow cells, the MKC of MPN patients were found to display a high level of BNIP-3 [50]. On the other hand, the most pronounced reductions in BNIP-3 were observed in PMF, suggesting a loss of proapoptotic potential during progression to the "accelerated" phase of MPNs [80].

JAK2 exon 12 mutations, exclusively associated with PV, consist of deletions/insertions, duplications, and point mutations, which affect a conserved region in the proximity of JAK2 pseudokinase domain (residues F537 through E543) and have functional consequences similar to those induced by *JAK2* V617F, however with some quantitative and qualitative differences [81]. Thus, exon 12 mutations cause a constitutive activation of EPO-R (**Figure 1a**) with erythroid hyperplasia, and lesser involvement of other hematopoietic lineages [82, 83]. Unlike *JAK2* V6217F, that is commonly homozygote in PV, at least in a proportion of colonies, exon 12 mutations are predominantly heterozygous, suggesting a stronger activation of JAK2 signaling [82, 84, 85]. As indicated by the levels of pAKT in the erythroid colonies, *JAK2* exon 12 mutations are associated with a weaker activation of AKT signaling compared to *JAK2* V617F. Both *JAK2* V617F and *JAK2* exon 12 mutations block the DNA damage-mediated apoptosis through inhibition of the BCL-xL deamidation pathway [81].

MPL exon 10 mutations induce an increased TPO-R signaling (**Figure 1d**) resulting in the activation of STAT5, STAT3, ERK, and AKT with associated thrombocytosis [86]. The most prevalent *MPL* mutations include substitutions of the juxtamembrane tryptophan W515, mainly by leucine (W515 L) or lysine (W515K) and rarely by alanine (W515A) or arginine (W515R). As tryptophan W515 is part of the amphipathic helical motif RWQFP that prevents TPO-R self-activation, these substitutions cause a cytokine-independent activation of the receptor [87, 88]. In this respect, in vitro expression studies of MPL mutants indicated that *MPL* W515K/L mutations were able to induce spontaneous cell proliferation and activation of JAK/STAT, RAS/MAPK, and PI3K/AKT pathways. In addition, an antiapoptotic effect was observed after cytokine withdrawal in MPL W515K/L-expressing cell lines, characterized by high levels of BCL-xL expression [89]. A second type of exon 10 *MPL* driver mutations, S505 N [52], was initially described as a germline mutation in a Japanese family suffering from hereditary thrombocytosis. Functional studies revealed that cell lines expressing S505 N presented growth factor-independent survival capacity accompanied by a constitutive phosphorylation of MEK1/2 and STAT5b. Furthermore, an autonomous phosphorylation of MEK1/2 was also noticed in the platelets obtained from the affected family members [90]. Later on, the same *MPL* mutation was reported as somatic in less than 1% ET cases [91].

Noncanonical *MPL* mutations (outside exon 10) of somatic or germline origin have been identified by whole-exome sequencing (WES) in approximately 10% of triple-negative ET or PMF cases [92]. By performing several functional assays, it was shown that all *MPL* mutations were gain-of-function resulting in activation of JAK2-STAT5 signaling. In a different study, the *MPL* somatic mutation S204P could induce TPO-independent growth, resistant to cytokine deprivation and constitutive STAT activation, although less efficient than *MPL* W515K mutation [93].

CALR mutations are located in the exon 9 of the gene and induce a + 1 base-pair frameshift, resulting in a new C-terminus that loses the ER retention motif (KDEL). Although more than 50 *CALR* mutations have been described so far, the mutation profile is dominated by a 52-bp deletion (type 1 mutation) and a 5-bp insertion (type 2) [94]. Functionally, both type 1 and type 2 CALR mutants bind to the extracellular domain of TPO-R and cause its activation (**Figure 1e**), leading to constitutive JAK2/STAT/PI-3 K and MAPK signaling and protecting cells from apoptosis [95]. Recently, by using gene expression analysis of K562 cells that lack TPO-R and stably express either CALR WT or the two most common CALR mutants, Salati et al. have identified a novel potential role of *CALR* mutations in MPN development, independent of TPO-R activation. Thus, CALR mutants seemed to diminish the proapoptotic signals downstream the unfolded protein response, generating the accumulation of misfolded proteins in ER and promoting resistance to ER stress-induced apoptosis [96].

Taking into account the above-mentioned cellular and molecular effects of MPN driver mutations, we can assume that megakaryocytic and erythroid progenitor expansion in MPNs results from a combination of increased proliferation and attenuated apoptosis.

5. PCD resistance in myeloid proliferation exploited in single or combined targeted therapeutic strategies

As shown previously, deregulation of the JAK/STAT pathway is central to MPN development and is driven in most cases by activating mutations in *JAK2*, *CALR*, or *MPL*. Signaling through other pathways (RAS/RAF/MEK/ERK, PI3K/AKT/mTOR, and LNK) and alterations in other cellular processes such as DNA methylation (e.g., *TET2* and *DNMT3A* mutations), histone modification (*ASXL1* and *EZH2* mutations), and RNA splicing (*U2AF1*, *SF3B1*, and *SRSF2* mutations) further contribute to initiation, progression of myeloproliferation, and resistance to apoptosis [52, 97].

Understanding molecular mechanisms of MPN pathogenesis has stimulated drug development in the field.

Reduction of thrombotic risk is the major goal of therapy in patients with PV and ET, and hydroxyurea (HU) is normally the first-line drug for achieving cytoreduction [98, 99]. In addition, most patients should receive aspirin, if they have no contraindications. In PV, maintaining hematocrit values <45% is an important therapeutic target. Second-line drugs of choice are interferon-α (IFNα) and busulfan [99].

The clinical efficacy of IFNα has been reported since 30 years ago and was improved with the development of pegylated forms [100, 101]. Furthermore, significant reductions of the *JAK2* V617F allele burden (% *JAK2* V617F) was observed in IFNα-treated patients [102, 103], suggesting that IFNα is able to target the malignant clone. The mechanism of action of IFNα in MPNs is not clearly elucidated, but several studies confirmed a targeted effect against *JAK2* V617F mutant clones. Ropeginterferon alpha-2b (Ropeg) is a long-acting pegylated-IFNα-2b, recently shown to be safe and well tolerated in phase 1–2 studies in PV patients. Both hematological and molecular responses have been reported in a phase 2 trial [104]. The discovery of *JAK2* V617F mutation and its role in constitutive activation of downstream signaling pathways and MPN pathogenesis triggered the search for specific JAK2 kinase inhibitors as potential targeted therapy. One of the first molecules approved for targeted treatment of MPN patients, ruxolitinib/jakafi (RUX), a selective JAK2 and JAK1 inhibitor, is currently used in PV resistant or intolerant to HU, and in intermediate and high risk PMF [99, 105]. RUX, approved in 2014 based on the results of the RESPONSE trial [106, 107], inhibits ATP-binding catalytic site of JAK kinase domain (both in mutant and wild type JAK), thus leading to a reduction of phosphorylation level of STAT-3/5, ERK, and AKT [108, 109]. Initial studies showed that RUX treatment of *JAK2* V617F-positive Ba/F3 cells inhibited cell proliferation and induced apoptosis [108]. The drug was able to cause apoptosis by decreasing BCL-xL, as well as proviral integrations of Moloney virus (PIM) 1 and 2 at transcriptional level, and consequently by inhibiting BAD phosphorylation [110].

RUX showed efficiency in spleen size reduction and symptomatology alleviation, improving quality of life, and overall survival; however, no significant decrease in allele burden was achieved [11, 111, 112]. RUX effects on the malignant clone are modest, side effects (such as anemia and thrombocytopenia) are reported, and drug resistance may appear. Other therapeutic strategies have been developed; they include the discovery of new inhibitors that target specifically mutant JAK2 and the combination of current therapies with other molecules that inhibit components of signaling pathway [105].

Early studies provided some evidence for the increased resistance to apoptosis of PV erythroid progenitor cells: overexpression of BCL-xL in the absence of EPO and a higher sensitivity to the antiapoptotic growth factor IGF-1 [113]. Moreover, Zeuner et al. have shown that erythroid precursors in PV patients with average and high *JAK2* V627F mutational load often expressed elevated levels of BCL-2 and BCL-xL and were very susceptible to the apoptosis induced by the BH3 mimetic ABT-737 (a small-molecule that inhibits BCL-2, BCL-xL, and BCL-W and causes apoptosis of the leukemic cells) compared to *JAK2* V617F-low or normal erythroblasts [114]. Later, the combination of ABT-737 with a JAK inhibitor proved to be efficient in reducing the number of PV *JAK2* V617F+EPO-dependent and independent erythroid colonies, and BIM was identified as a key mediator of apoptosis induced by JAK2 inhibition [115].

In susceptible cells, apoptosis is caused by exposure to a JAK inhibitor, which leads to dephosphorylation of BAD, enabling BAD to bind and sequester the anti-apoptotic protein BCL-xL. On the other side, in potent cells, RAS effector pathways keep BAD phosphorylation in the presence of JAK inhibitors, maintaining a specific dependence on BCL-xL for survival. So, downstream regulation of BCL-xL, more precisely BCL-xL inhibition, might be the key against resistance to JAK inhibition by either co-inhibition of JAK and RAS effector in AKT and ERK pathways or by direct inhibition of BCL-xL inducing apoptosis [116].

At present, there are over 1500 clinical trials evaluating various drug effects on myeloproliferative neoplasms registered at clinicaltrials.gov. Some of them might be successful due to targeting different apoptotic pathways or by targeting simultaneously different types of PCD.

Plitidepsin is a synthetically produced anticancer agent [117], a cyclodepsipeptide related to didemnins, commercialized as Aplidin® (PharmaMar, S.A., Madrid, Spain). Plitidepsin induces dose-dependent cell-cycle arrest and an acute apoptotic process. These effects rely on the induction of early oxidative stress, the rapid activation of Rac1 GTPase, and the sustained activation of JNK and p38/MAPK, which finally result in caspase-dependent apoptosis [118, 119]. JNK phosphorylation can be seen as early as 5–10 minutes after exposure to the compound. The activation of JNK and p38/MAPK is associated with increase in reactive oxygen species and a decrease in reduced form of glutathione [120].

Recent studies have led researchers to hypothesize that the primary target of plitidepsin could be the eukaryotic elongation factor 1A2 (eEF1A2), which is overexpressed in tumors and supports tumor cell proliferation while inhibiting apoptosis [121]. eEF1A2 seems to be an interesting target for therapy and may also be a biomarker predicting drug sensitivity. Aplidin®/Plitidepsin was investigated for its effect (safety and tolerability) on bone marrow or peripheral blood cells as well as assessed the response rate in patients with PMF, post-PV MF, or post-ET MF, in phase II/open label single agent clinical trial (NCT01149681). Although the drug was well tolerated, the trial was prematurely terminated due to the low response rate [122].

Navitoclax is an orally active, synthetic small molecule and an antagonist of the apoptosis suppressor proteins BCL-2, BCL-xL, and BCL-w, which are frequently overexpressed in a wide variety of cancers, including myeloid ones, and are linked to drug resistance. Inhibition of these apoptosis suppressors prevents their binding to the apoptotic effectors BAX and BAK proteins, thereby triggering apoptotic processes in cells overexpressing BCL-2, BCL-xL, and BCL-w. This eventually reduces tumor cell proliferation. Navitoclax (ABT-263) and RUX are currently evaluated in combination for efficacy, safety, and tolerability on spleen volume as assessed by magnetic resonance imaging (MRI) in participants with MF in a phase II/open label clinical trial (NCT03222609).

Obatoclax (GX15–070) is a BH3-mimetic designed to target and counteract anti-apoptotic BCL-2 proteins. Obatoclax is an MCL-1 antagonist [123] and downregulates p53, and it has a dual mechanism of action, being capable to induce apoptosis or autophagy [124]. On the other side, obatoclax accumulates in lysosomes inducing their alkalinization and inhibiting their function [125]. Parikh et al. conducted a multicenter, open-label, noncomparative phase II study (NCT00360035) of obatoclax in patients with MF. Unfortunately, obatoclax exhibited no significant clinical activity at the tested dose and schedule [126].

Other phase I trial (NCT02436135) investigated the combination of RUX with idelalisib, a PI3K delta inhibitor, as therapy for intermediate to high-risk PMF, post-PV MF, or post-ET MF with progressive or relapsed disease [127].

PIM inhibitors have shown preclinical synergy with JAK inhibitors, as well as the ability to overcome JAK inhibitor resistance in MPN cell lines. PIM regulate JAK/STAT signaling and are involved in oncogenesis through phosphorylation of cell cycle regulators, activation of antiapoptotic proteins, and enhancement of MYC expression [97]. A phase 1b study of RUX plus PIM inhibitor PIM447, or RUX plus CDK4/6 inhibitor ribociclib (LEE011), or the combination of all three is underway in several non-U.S. countries (NCT02370706).

As PI3K/AKT/mTOR signaling is markedly activated in MPNs, small molecule inhibitors of the proteins involved in this pathway have been tested in MF with promising results. Thus, mTOR inhibitor everolimus, as single therapeutic agent, was able to induce responses, in terms of reducing constitutional symptoms and the degree of leukocytosis, thrombocytosis, and anemia, in 23% of patients in a phase I/II clinical trial. Due to the preclinically proved synergic effects of PI3K/AKT/mTOR inhibitors and JAK inhibitors, several clinical studies were initiated: PI3K inhibitor TGR-1202 in combination with RUX (NCT02493530), PI3K inhibitor buparlisib with RUX (NCT01730248), PI3K inhibitor INCB050465 and RUX (NCT02718300), and selective PI3Kδ inhibitor TGR-1202 and RUX (NCT02493530). Preliminary results of buparlisib and RUX phase 1b study indicated that this drug association was well tolerated, and $\geq$ 50% reduction in splenomegaly was observed in 70% of JAK-inhibitor naive patients and 54% of patients who did not previously respond to JAK2 inhibitor monotherapy [97, 127].

RAF/MEK/ERK pathway is another signaling cascade activated in MPNs by the increased JAK/STAT signaling. Therefore, MEK inhibitors were tested in different murine models, either alone or in combination with JAK inhibitors, showing a decrease in bone marrow fibrosis, inhibition of malignant cell growth, and HSC function recovery, associated with a prolonged survival. Moreover, a new trial that combines the MEK inhibitor selumetinib with the DNA hypomethylating agent azacitidine is soon expected [97].

Preclinical studies reveal a central role for tumor necrosis factor alfa (TNF-α) in promoting clonal dominance of *JAK2* V617F-expressing cells in MPN [101]. *JAK2* V617F appears to confer TNF-α resistance to a preneoplastic TNF-α-sensitive cell, while creating a TNF-α-rich environment at the same time.

SMAC-mimetics are novel apoptosis-inducing agents that stimulate the ubiquitinylation and proteasomal degradation of cellular inhibitors of apoptosis (IAPs) [102], proteins that play an important role in tumor cell resistance to cytotoxicity mediated by TNF superfamily cytokines. These agents have been shown to sensitize cancer cells to TNF family-induced apoptosis [103]. Results from a phase II trial of the SMAC-mimetic LCL-161 in patients with intermediate or high-risk MF intolerant of, ineligible for, or relapsed/refractory to JAK inhibitors were recently presented. Six of sixteen evaluated patients (38%) had objective responses, obtaining clinical improvement and in one case cytogenetic remission [128].

Resistance of hematologic malignancies to PCD significantly limits the efficacy of chemotherapy. As the majority of chemotherapeutic drugs trigger apoptosis, the observed resistance may indicate that novel therapeutic strategies to reactivate nonapoptotic PCD or at least combined therapeutic strategies able to attack simultaneously different mechanisms might be better approaches to eradicate malignant cells.

6. Conclusion

Deregulation of pro- and anti-PCD genes involved in cell resistance to cell death and accumulation of myeloid cells in MPNs is continuously clarified by intense exploration of the modifications affecting different types of PCD pathways involved in myeloid proliferation. At the same time, comprehension of the network of signaling pathways involved in etiology and drug resistance of these disorders facilitate a more efficient exploitation of the knowledge, using combined and synergic, targeted therapeutic strategies.

Acknowledgements

We gratefully acknowledge the funding from the project Competitiveness Operational Programme (COP) A1.1.4. ID: P_37_798 MyeloAL-EDiaProT, Contract 149/26.10.2016, (SMIS: 106774), MyeloAL Project.

Notes

All the authors contributed equally to this manuscript.

Author details

Carmen C. Diaconu[1*], Petruta Gurban[1,2], Cristina Mambet[1], Mihaela Chivu-Economescu[1], Laura G. Necula[1,3], Lilia Matei[1], Denisa Dragu[1], Saviana Nedeianu[1], Ana I. Neagu[1], Aurelia Tatic[1,4,5], Diana Cristodor[1] and Coralia Bleotu[1]

1 Stefan S. Nicolau Institute of Virology, Bucharest, Romania

2 Personal Genetics Ltd., Bucharest, Romania

3 Titu Maiorescu University, Bucharest, Romania

4 Carol Davila University of Medicine and Pharmacy, Bucharest, Romania

5 Center of Hematology and Bone Marrow Transplantation, Fundeni Clinical Institute, Bucharest, Romania

*Address all correspondence to: ccdiaconu@yahoo.com

References

[1] Orkin SH, Zon LI. Hematopoiesis: An evolving paradigm for stem cell biology. Cell. 2008;**132**(4):631-644. DOI: 10.1016/j.cell.2008.01.025

[2] Sarvothaman S, Undi RB, Pasupuleti SR, Gutti U, Gutti RK. Apoptosis: Role in myeloid cell development. Blood Research. 2015;**50**(2):73-79. DOI: 10.5045/br.2015.50.2.73

[3] Ishida S, Akiyama H, Umezawa Y, Okada K, Nogami A, Oshikawa G, et al. Mechanisms for mTORC1 activation and synergistic induction of apoptosis by ruxolitinib and BH3 mimetics or autophagy inhibitors in JAK2-V617F-expressing leukemic cells including newly established PVTL-2. Oncotarget. 2018;**9**(42):26834-26851. DOI: 10.18632/oncotarget.25515

[4] Murphy JM, Silke J. Ars Moriendi; the art of dying well—New insights into the molecular pathways of necroptotic cell death. EMBO Reports. 2014;**15**(2):155-164. DOI: 10.1002/embr.201337970

[5] Spivak JL. Myeloproliferative neoplasms. The New England Journal of Medicine. 2017;**376**(22):2168-2181. DOI: 10.1056/NEJMra1406186

[6] Opferman JT, Kothari A. Anti-apoptotic BCL-2 family members in development. Cell Death and Differentiation. 2018;**25**(1):37-45. DOI: 10.1038/cdd.2017.170

[7] McArthur K, Chappaz S, Kile BT. Apoptosis in megakaryocytes and platelets: The life and death of a lineage. Blood. 2018;**131**(6):605-610. DOI: 10.1182/blood-2017-11-742684

[8] Gores GJ, Kaufmann SH. Selectively targeting Mcl-1 for the treatment of acute myelogenous leukemia and solid tumors. Genes & Development. 2012;**26**(4):305-311. DOI: 10.1101/gad.186189.111

[9] Kile BT. The role of apoptosis in megakaryocytes and platelets. British Journal of Haematology. 2014;**165**(2):217-226. DOI: 10.1111/bjh.12757

[10] Testa U. Apoptotic mechanisms in the control of erythropoiesis. Leukemia. 2004;**18**(7):1176-1199. DOI: 10.1038/sj.leu.2403383

[11] Tognon R, Nunes Nde S, Castro FA. Apoptosis deregulation in myeloproliferative neoplasms. Einstein (Sao Paulo). 2013;**11**(4):540-544

[12] Thomas MP, Liu X, Whangbo J, McCrossan G, Sanborn KB, Basar E, et al. Apoptosis triggers specific, rapid, and global mRNA decay with 3' uridylated intermediates degraded by DIS3L2. Cell Reports. 2015;**11**(7):1079-1089. DOI: 10.1016/j.celrep.2015.04.026

[13] Craig RW. MCL1 provides a window on the role of the BCL2 family in cell proliferation, differentiation and tumorigenesis. Leukemia. 2002;**16**(4):444-454. DOI: 10.1038/sj.leu.2402416

[14] Opferman JT. Life and death during hematopoietic differentiation. Current Opinion in Immunology. 2007;**19**(5):497-502. DOI: 10.1016/j.coi.2007.06.002

[15] Kodama T, Hikita H, Kawaguchi T, Shigekawa M, Shimizu S, Hayashi Y, et al. Mcl-1 and Bcl-xL regulate Bak/Bax-dependent apoptosis of the megakaryocytic lineage at multistages. Cell Death and Differentiation. 2012;**19**(11):1856-1869. DOI: 10.1038/cdd.2012.88

[16] Kollek M, Muller A, Egle A, Erlacher M. Bcl-2 proteins in development, health, and disease of the hematopoietic system. The FEBS Journal. 2016;**283**(15):2779-2810. DOI: 10.1111/febs.13683

[17] Mizrahi K, Askenasy N. Physiological functions of TNF family receptor/ligand interactions in hematopoiesis and transplantation. Blood. 2014;**124**(2):176-183. DOI: 10.1182/blood-2014-03-559641

[18] Rumi E, Cazzola M. Diagnosis, risk stratification, and response evaluation in classical myeloproliferative neoplasms. Blood. 2017;**129**(6):680-692. DOI: 10.1182/blood-2016-10-695957

[19] Pellagatti A, Vetrie D, Langford CF, Gama S, Eagleton H, Wainscoat JS, et al. Gene expression profiling in polycythemia vera using cDNA microarray technology. Cancer Research. 2003;**63**(14):3940-3944

[20] Gallardo M, Barrio S, Fernandez M, Paradela A, Arenas A, Toldos O, et al. Proteomic analysis reveals heat shock protein 70 has a key role in polycythemia Vera. Molecular Cancer. 2013;**12**:142. DOI: 10.1186/1476-4598-12-142

[21] Zeuner A, Pedini F, Signore M, Ruscio G, Messina C, Tafuri A, et al. Increased death receptor resistance and FLIPshort expression in polycythemia vera erythroid precursor cells. Blood. 2006;**107**(9):3495-3502. DOI: 10.1182/blood-2005-07-3037

[22] Tognon R, Gasparotto EP, Leroy JM, Oliveira GL, Neves RP, Carrara Rde C, et al. Differential expression of apoptosis-related genes from death receptor pathway in chronic myeloproliferative diseases. Journal of Clinical Pathology. 2011;**64**(1):75-82. DOI: 10.1136/jcp.2010.080895

[23] Tenedini E, Fagioli ME, Vianelli N, Tazzari PL, Ricci F, Tagliafico E, et al. Gene expression profiling of normal and malignant CD34-derived megakaryocytic cells. Blood. 2004;**104**(10):3126-3135. DOI: 10.1182/blood-2003-07-2597

[24] Florena AM, Tripodo C, Di Bernardo A, Iannitto E, Guarnotta C, Porcasi R, et al. Different immunophenotypical apoptotic profiles characterise megakaryocytes of essential thrombocythaemia and primary myelofibrosis. Journal of Clinical Pathology. 2009;**62**(4):331-338. DOI: 10.1136/jcp.2007.054353

[25] Trelinski J, Chojnowski K, Cebula-Obrzut B, Smolewski P. Impaired apoptosis of megakaryocytes and bone marrow mononuclear cells in essential thrombocythemia: Correlation with JAK2V617F mutational status and cytoreductive therapy. Medical Oncology. 2012;**29**(4):2388-2395. DOI: 10.1007/s12032-012-0202-3

[26] Tefferi A. Primary myelofibrosis: 2014 update on diagnosis, risk-stratification, and management. American Journal of Hematology. 2014;**89**(9):915-925. DOI: 10.1002/ajh.23703

[27] Cahu X, Constantinescu SN. Oncogenic drivers in myeloproliferative neoplasms: From JAK2 to calreticulin mutations. Current Hematologic Malignancy Reports. 2015;**10**(4):335-343. DOI: 10.1007/s11899-015-0278-x

[28] Thiele J, Lorenzen J, Manich B, Kvasnicka HM, Zirbes TK, Fischer R. Apoptosis (programmed cell death) in idiopathic (primary) osteo−/myelofibrosis: Naked nuclei in megakaryopoiesis reveal features of para-apoptosis. Acta Haematologica. 1997;**97**(3):137-143. DOI: 10.1159/000203671

[29] Ciurea SO, Merchant D, Mahmud N, Ishii T, Zhao Y, Hu W, et al. Pivotal contributions of megakaryocytes to the biology of idiopathic myelofibrosis. Blood. 2007;**110**(3):986-993. DOI: 10.1182/blood-2006-12-064626

[30] Theophile K, Hussein K, Kreipe H, Bock O. Expression

profiling of apoptosis-related genes in megakaryocytes: BNIP3 is downregulated in primary myelofibrosis. Experimental Hematology. 2008;**36**(12):1728-1738. DOI: 10.1016/j.exphem.2008.07.011

[31] Kleppe M, Kwak M, Koppikar P, Riester M, Keller M, Bastian L, et al. JAK–STAT Pathway Activation in Malignant and Nonmalignant Cells Contributes to MPN Pathogenesis and Therapeutic Response. 2015;**5**(3): 316-331. DOI:10.1158/2159-8290.CD-14-0736%J Cancer Discovery

[32] Desterke C, Martinaud C, Ruzehaji N, Le Bousse-Kerdiles MC. Inflammation as a keystone of bone marrow stroma alterations in primary myelofibrosis. Mediators of Inflammation. 2015;**2015**:415024. DOI: 10.1155/2015/415024

[33] Heaton WL, Senina AV, Pomicter AD, Salama ME, Clair PM, Yan D, et al. Autocrine Tnf signaling favors malignant cells in myelofibrosis in a Tnfr2-dependent fashion. Leukemia. 2018;**32**:2399-2411. DOI: 10.1038/s41375-018-0131-z

[34] Long JS, Ryan KM. New frontiers in promoting tumour cell death: Targeting apoptosis, necroptosis and autophagy. Oncogene. 2012;**31**(49):5045-5060. DOI: 10.1038/onc.2012.7

[35] Huang X, Xiao F, Li Y, Qian W, Ding W, Ye X. Bypassing drug resistance by triggering necroptosis: Recent advances in mechanisms and its therapeutic exploitation in leukemia. Journal of Experimental & Clinical Cancer Research. 2018;**37**(1):310. DOI: 10.1186/s13046-018-0976-z

[36] Hardwick JM, Soane L. Multiple functions of BCL-2 family proteins. Cold Spring Harbor Perspectives in Biology. 2013;**5**(2):1-22. DOI: 10.1101/cshperspect.a008722

[37] Visconte V, Przychodzen B, Han Y, Nawrocki ST, Thota S, Kelly KR, et al. Complete mutational spectrum of the autophagy interactome: A novel class of tumor suppressor genes in myeloid neoplasms. Leukemia. 2017;**31**(2): 505-510. DOI: 10.1038/leu.2016.295

[38] Sehgal AR, Konig H, Johnson DE, Tang D, Amaravadi RK, Boyiadzis M, et al. You eat what you are: autophagy inhibition as a therapeutic strategy in leukemia. Leukemia. 2015;**29**(3):517-525. DOI: 10.1038/leu.2014.349

[39] Adams JM, Cory S. The BCL-2 arbiters of apoptosis and their growing role as cancer targets. Cell Death and Differentiation. 2018;**25**(1):27-36. DOI: 10.1038/cdd.2017.161

[40] White MJ, Schoenwaelder SM, Josefsson EC, Jarman KE, Henley KJ, James C, et al. Caspase-9 mediates the apoptotic death of megakaryocytes and platelets, but is dispensable for their generation and function. Blood. 2012;**119**(18):4283-4290. DOI: 10.1182/blood-2011-11-394858

[41] Malherbe JA, Fuller KA, Mirzai B, Kavanagh S, So CC, Ip HW, et al. Dysregulation of the intrinsic apoptotic pathway mediates megakaryocytic hyperplasia in myeloproliferative neoplasms. Journal of Clinical Pathology. 2016;**69**(11):1017-1024. DOI: 10.1136/jclinpath-2016-203625

[42] Zhang J, Huang K, O'Neill KL, Pang X, Luo X. Bax/Bak activation in the absence of Bid, Bim, Puma, and p53. Cell Death & Disease. 2016;**7**:e2266. DOI: 10.1038/cddis.2016.167

[43] Fletcher JI, Meusburger S, Hawkins CJ, Riglar DT, Lee EF, Fairlie WD, et al. Apoptosis is triggered when prosurvival Bcl-2 proteins cannot restrain Bax. Proceedings of the National Academy of Sciences of the United States of

America. 2008;**105**(47):18081-18087. DOI: 10.1073/pnas.0808691105

[44] Tognon R, Gasparotto EP, Neves RP, Nunes NS, Ferreira AF, Palma PV, et al. Deregulation of apoptosis-related genes is associated with PRV1 overexpression and JAK2 V617F allele burden in essential thrombocythemia and myelofibrosis. Journal of Hematology & Oncology. 2012;**5**:2. DOI: 10.1186/1756-8722-5-2

[45] Gasparotto EPL, Tognon R, Ferreira AF, Oliveira GLV, Palma PVB, Zanichelli MA, et al. Deregulated expression of A1, Bcl-2, Bcl-xL, and Mcl-1 antiapoptotic proteins and Bid, Bad, and Bax proapoptotic genes in polycythemia vera patients. Brazilian Journal of Pharmaceutical Sciences. 2011;**47**:873-886

[46] Rubert J, Qian Z, Andraos R, Guthy DA, Radimerski T. Bim and Mcl-1 exert key roles in regulating JAK2V617F cell survival. BMC Cancer. 2011;**11**:24. DOI: 10.1186/1471-2407-11-24

[47] Zhao R, Follows GA, Beer PA, Scott LM, Huntly BJ, Green AR, et al. Inhibition of the Bcl-xL deamidation pathway in myeloproliferative disorders. The New England Journal of Medicine. 2008;**359**(26):2778-2789. DOI: 10.1056/NEJMoa0804953

[48] Malherbe JA, Fuller KA, Arshad A, Nangalia J, Romeo G, Hall SL, et al. Megakaryocytic hyperplasia in myeloproliferative neoplasms is driven by disordered proliferative, apoptotic and epigenetic mechanisms. Journal of Clinical Pathology. 2016;**69**(2):155-163. DOI: 10.1136/jclinpath-2015-203177

[49] Humphry NJ, Wheatley SP. Survivin inhibits excessive autophagy in cancer cells but does so independently of its interaction with LC3. Biology Open. 2018;**7**(10):1-7. DOI: 10.1242/bio.037374

[50] Koopmans SM, Schouten HC, van Marion AM. Anti-apoptotic pathways in bone marrow and megakaryocytes in myeloproliferative neoplasia. Pathobiology. 2014;**81**(2):60-68. DOI: 10.1159/000356187

[51] Giatromanolaki A, Koukourakis MI, Sowter HM, Sivridis E, Gibson S, Gatter KC, et al. BNIP3 expression is linked with hypoxia-regulated protein expression and with poor prognosis in non-small cell lung cancer. Clinical Cancer Research. 2004;**10**(16):5566-5571. DOI: 10.1158/1078-0432.CCR-04-0076

[52] Vainchenker W, Kralovics R. Genetic basis and molecular pathophysiology of classical myeloproliferative neoplasms. Blood. 2017;**129**(6):667-679. DOI: 10.1182/blood-2016-10-695940

[53] Wang WA, Groenendyk J, Michalak M. Calreticulin signaling in health and disease. The International Journal of Biochemistry & Cell Biology. 2012;**44**(6):842-846. DOI: 10.1016/j.biocel.2012.02.009

[54] Mondet J, Park JH, Menard A, Marzac C, Carillo S, Pourcelot E, et al. Endogenous megakaryocytic colonies underline association between megakaryocytes and calreticulin mutations in essential thrombocythemia. Haematologica. 2015;**100**(5):e176-e178. DOI: 10.3324/haematol.2014.118927

[55] Kominami K, Nakabayashi J, Nagai T, Tsujimura Y, Chiba K, Kimura H, et al. The molecular mechanism of apoptosis upon caspase-8 activation: quantitative experimental validation of a mathematical model. Biochimica et Biophysica Acta. 2012;**1823**(10):1825-1840. DOI: 10.1016/j.bbamcr.2012.07.003

[56] Straathof KC, Pule MA, Yotnda P, Dotti G, Vanin EF, Brenner MK, et al.

An inducible caspase 9 safety switch for T-cell therapy. Blood. 2005;**105**(11). DOI: 4247-4254 DOI:10.1182/blood-2004-11-4564

[57] Paul A, Krelin Y, Arif T, Jeger R, Shoshan-Barmatz V. A new role for the mitochondrial pro-apoptotic protein SMAC/Diablo in phospholipid synthesis associated with tumorigenesis. Molecular Therapy. 2018;**26**(3):680-694. DOI: 10.1016/j.ymthe.2017.12.020

[58] Song H, Hollstein M, Xu Y. p53 gain-of-function cancer mutants induce genetic instability by inactivating ATM. Nature Cell Biology. 2007;**9**(5):573-580. DOI: 10.1038/ncb1571

[59] Chen S, Gao R, Yao C, Kobayashi M, Liu SZ, Yoder MC, et al. Genotoxic stresses promote clonal expansion of hematopoietic stem cells expressing mutant p53. Leukemia. 2018;**32**(3):850-854. DOI: 10.1038/leu.2017.325

[60] Hanson B. Necroptosis: A new way of dying? Cancer Biology & Therapy. 2016;**17**(9):899-910. DOI: 10.1080/15384047.2016.1210732

[61] Diaconu C, Mambet C, Necula LG, Gurban P, Matei L, Aldea-Pitica IM, et al. Triple negative myeloproliferative neoplasms-sometimes driver mutations stay low-key in plain sight. Romanian Biotechnological Letters. 2018;**23**(4):13750-13760. DOI: 10.26327/RBL2017.12

[62] Mambet C, Matei L, Necula LG, Diaconu CC. A link between the driver mutations and dysregulated apoptosis in BCR-ABL1 negative myeloproliferative neoplasms. Journal of Immunoassay & Immunochemistry. 2016;**37**(4):331-345. DOI: 10.1080/15321819.2016.1152276

[63] Vainchenker W, Leroy E, Gilles L, Marty C, Plo I, Constantinescu SN. JAK inhibitors for the treatment of myeloproliferative neoplasms and other disorders. F1000Res. 2018;7:82. DOI: 10.12688/f1000research.13167.1

[64] James C, Ugo V, Le Couedic JP, Staerk J, Delhommeau F, Lacout C, et al. A unique clonal JAK2 mutation leading to constitutive signalling causes polycythaemia vera. Nature. 2005;**434**(7037):1144-1148. DOI: 10.1038/nature03546

[65] Kralovics R, Passamonti F, Buser AS, Teo SS, Tiedt R, Passweg JR, et al. A gain-of-function mutation of JAK2 in myeloproliferative disorders. The New England Journal of Medicine. 2005;**352**(17):1779-1790. DOI: 10.1056/NEJMoa051113

[66] Levine RL, Wadleigh M, Cools J, Ebert BL, Wernig G, Huntly BJ, et al. Activating mutation in the tyrosine kinase JAK2 in polycythemia vera, essential thrombocythemia, and myeloid metaplasia with myelofibrosis. Cancer Cell. 2005;**7**(4):387-397. DOI: 10.1016/j.ccr.2005.03.023

[67] Mustjoki S, Borze I, Lasho TL, Alitalo R, Pardanani A, Knuutila S, et al. JAK2V617F mutation and spontaneous megakaryocytic or erythroid colony formation in patients with essential thrombocythaemia (ET) or polycythaemia vera (PV). Leukemia Research. 2009;**33**(1):54-59. DOI: 10.1016/j.leukres.2008.07.008

[68] Dusa A, Staerk J, Elliott J, Pecquet C, Poirel HA, Johnston JA, et al. Substitution of pseudokinase domain residue Val-617 by large non-polar amino acids causes activation of JAK2. The Journal of Biological Chemistry. 2008;**283**(19):12941-12948. DOI: 10.1074/jbc.M709302200

[69] Yasuda M, Han JW, Dionne CA, Boyd JM, Chinnadurai G. BNIP3alpha: a human homolog of mitochondrial

proapoptotic protein BNIP3. Cancer Research. 1999;**59**(3):533-537

[70] Azad MB, Chen Y, Henson ES, Cizeau J, McMillan-Ward E, Israels SJ, et al. Hypoxia induces autophagic cell death in apoptosis-competent cells through a mechanism involving BNIP3. Autophagy. 2008;**4**(2):195-204

[71] Rouyez MC, Boucheron C, Gisselbrecht S, Dusanter-Fourt I, Porteu F. Control of thrombopoietin-induced megakaryocytic differentiation by the mitogen-activated protein kinase pathway. Molecular and Cellular Biology. 1997;**17**(9):4991-5000

[72] Morrison DK. MAP kinase pathways. Cold Spring Harbor Perspectives in Biology. 2012;**4**(11). DOI: 10.1101/cshperspect.a011254

[73] Laubach JP, Fu P, Jiang X, Salter KH, Potti A, Arcasoy MO. Polycythemia vera erythroid precursors exhibit increased proliferation and apoptosis resistance associated with abnormal RAS and PI3K pathway activation. Experimental Hematology. 2009;**37**(12):1411-1422. DOI: 10.1016/j.exphem.2009.09.009

[74] Park S, Chapuis N, Tamburini J, Bardet V, Cornillet-Lefebvre P, Willems L, et al. Role of the PI3K/AKT and mTOR signaling pathways in acute myeloid leukemia. Haematologica. 2010;**95**(5):819-828. DOI: 10.3324/haematol.2009.013797

[75] Fransecky L, Mochmann LH, Baldus CD. Outlook on PI3K/AKT/mTOR inhibition in acute leukemia. Molecular and Cellular Therapies. 2015;**3**:2. DOI: 10.1186/s40591-015-0040-8

[76] Dai C, Chung IJ, Krantz SB. Increased erythropoiesis in polycythemia vera is associated with increased erythroid progenitor proliferation and increased phosphorylation of Akt/PKB. Experimental Hematology. 2005;**33**(2):152-158. DOI: 10.1016/j.exphem.2004.10.017

[77] Khan I, Huang Z, Wen Q, Stankiewicz MJ, Gilles L, Goldenson B, et al. AKT is a therapeutic target in myeloproliferative neoplasms. Leukemia. 2013;**27**(9):1882-1890. DOI: 10.1038/leu.2013.167

[78] Fu C, Wen QJ, Marinaccio C, Ling T, Chen W, Bulic M, et al. AKT activation is a feature of CALR mutant myeloproliferative neoplasms. Leukemia. 2019;**33**(1):271-274. DOI: 10.1038/s41375-018-0224-8

[79] Bartalucci N, Tozzi L, Bogani C, Martinelli S, Rotunno G, Villeval JL, et al. Co-targeting the PI3K/mTOR and JAK2 signalling pathways produces synergistic activity against myeloproliferative neoplasms. Journal of Cellular and Molecular Medicine. 2013;**17**(11):1385-1396. DOI: 10.1111/jcmm.12162

[80] Tefferi A. Myeloproliferative neoplasms: Thrombophilic clonal stem cell diseases. Cancer Treatment and Research. 2009;**148**:157-179. DOI: 10.1007/978-0-387-79962-9_10

[81] Scott LM. The JAK2 exon 12 mutations: A comprehensive review. American Journal of Hematology. 2011;**86**(8):668-676. DOI: 10.1002/ajh.22063

[82] Scott LM, Tong W, Levine RL, Scott MA, Beer PA, Stratton MR, et al. JAK2 exon 12 mutations in polycythemia vera and idiopathic erythrocytosis. The New England Journal of Medicine. 2007;**356**(5):459-468. DOI: 10.1056/NEJMoa065202

[83] Pardanani A, Lasho TL, Finke C, Hanson CA, Tefferi A. Prevalence and clinicopathologic correlates of JAK2 exon 12 mutations in

JAK2V617F-negative polycythemia vera. Leukemia. 2007;**21**(9):1960-1963. DOI: 10.1038/sj.leu.2404810

[84] Butcher CM, Hahn U, To LB, Gecz J, Wilkins EJ, Scott HS, et al. Two novel JAK2 exon 12 mutations in JAK2V617F-negative polycythaemia vera patients. Leukemia. 2008;**22**(4):870-873. DOI: 10.1038/sj.leu.2404971

[85] Schnittger S, Bacher U, Haferlach C, Geer T, Muller P, Mittermuller J, et al. Detection of JAK2 exon 12 mutations in 15 patients with JAK2V617F negative polycythemia vera. Haematologica. 2009;**94**(3):414-418. DOI: 10.3324/haematol.13223

[86] Grinfeld J, Nangalia J, Green AR. Molecular determinants of pathogenesis and clinical phenotype in myeloproliferative neoplasms. Haematologica. 2017;**102**(1):7-17. DOI: 10.3324/haematol.2014.113845

[87] Staerk J, Lacout C, Sato T, Smith SO, Vainchenker W, Constantinescu SN. An amphipathic motif at the transmembrane-cytoplasmic junction prevents autonomous activation of the thrombopoietin receptor. Blood. 2006;**107**(5):1864-1871. DOI:10.1182/blood-2005-06-2600

[88] Defour JP, Itaya M, Gryshkova V, Brett IC, Pecquet C, Sato T, et al. Tryptophan at the transmembrane-cytosolic junction modulates thrombopoietin receptor dimerization and activation. Proceedings of the National Academy of Sciences of the United States of America. 2013;**110**(7):2540-2545. DOI: 10.1073/pnas.1211560110

[89] Chaligne R, Tonetti C, Besancenot R, Roy L, Marty C, Mossuz P, et al. New mutations of MPL in primitive myelofibrosis: Only the MPL W515 mutations promote a G1/S-phase transition. Leukemia. 2008;**22**(8):1557-1566. DOI: 10.1038/leu.2008.137

[90] Ding J, Komatsu H, Wakita A, Kato-Uranishi M, Ito M, Satoh A, et al. Familial essential thrombocythemia associated with a dominant-positive activating mutation of the c-MPL gene, which encodes for the receptor for thrombopoietin. Blood. 2004;**103**(11):4198-4200. DOI: 10.1182/blood-2003-10-3471

[91] Beer PA, Campbell PJ, Scott LM, Bench AJ, Erber WN, Bareford D, et al. MPL mutations in myeloproliferative disorders: Analysis of the PT-1 cohort. Blood. 2008;**112**(1):141-149. DOI: 10.1182/blood-2008-01-131664

[92] Milosevic Feenstra JD, Nivarthi H, Gisslinger H, Leroy E, Rumi E, Chachoua I, et al. Whole-exome sequencing identifies novel MPL and JAK2 mutations in triple-negative myeloproliferative neoplasms. Blood. 2016;**127**(3):325-332. DOI: 10.1182/blood-2015-07-661835

[93] Cabagnols X, Favale F, Pasquier F, Messaoudi K, Defour JP, Ianotto JC, et al. Presence of atypical thrombopoietin receptor (MPL) mutations in triple-negative essential thrombocythemia patients. Blood. 2016;**127**(3):333-342. DOI: 10.1182/blood-2015-07-661983

[94] Cazzola M, Kralovics R. From Janus kinase 2 to calreticulin: The clinically relevant genomic landscape of myeloproliferative neoplasms. Blood. 2014;**123**(24):3714-3719. DOI: 10.1182/blood-2014-03-530865

[95] Chachoua I, Pecquet C, El-Khoury M, Nivarthi H, Albu RI, Marty C, et al. Thrombopoietin receptor activation by myeloproliferative neoplasm associated calreticulin mutants. Blood. 2016;**127**(10):1325-1335. DOI: 10.1182/blood-2015-11-681932

[96] Salati S, Genovese E, Prudente Z, Carretta C, Bartalucci N, Pennucci V, et al. Calreticulin Ins5 and Del52

mutations impair unfolded protein and oxidative stress responses in hematopoietic cells. American Society of Hematology. 2018

[97] Pettit K, Odenike O. Novel therapies for myelofibrosis. Current Hematologic Malignancy Reports. 2017;**12**(6):611-624. DOI: 10.1007/s11899-017-0403-0

[98] Tefferi A, Vannucchi AM, Barbui T. Polycythemia vera treatment algorithm 2018. Blood Cancer Journal. 2018;**8**(1):1-7. DOI:10.1038/s41408-017-0042-7

[99] Tefferi A, Barbui T. Polycythemia vera and essential thrombocythemia: 2019 update on diagnosis, risk-stratification and management. American Journal of Hematology. 2019;**94**(1):133-143. DOI: 10.1002/ajh.25303

[100] Marchioli R, Finazzi G, Specchia G, Cacciola R, Cavazzina R, Cilloni D, et al. Cardiovascular events and intensity of treatment in polycythemia vera. The New England Journal of Medicine. 2013;**368**(1):22-33. DOI: 10.1056/NEJMoa1208500

[101] Fleischman AG, Aichberger KJ, Luty SB, Bumm TG, Petersen CL, Doratotaj S, et al. TNFalpha facilitates clonal expansion of JAK2V617F positive cells in myeloproliferative neoplasms. Blood. 2011;**118**(24):6392-6398. DOI: 10.1182/blood-2011-04-348144

[102] Darding M, Feltham R, Tenev T, Bianchi K, Benetatos C, Silke J, et al. Molecular determinants of Smac mimetic induced degradation of cIAP1 and cIAP2. Cell Death and Differentiation. 2011;**18**(8):1376-1386. DOI: 10.1038/cdd.2011.10

[103] Welsh K, Milutinovic S, Ardecky RJ, Gonzalez-Lopez M, Ganji SR, Teriete P, et al. Characterization of potent SMAC mimetics that sensitize cancer cells to TNF family-induced apoptosis. PLoS One. 2016;**11**(9):e0161952:1-19. DOI: 10.1371/journal.pone.0161952

[104] Verger E, Soret-Dulphy J, Maslah N, Roy L, Rey J, Ghrieb Z, et al. Ropeginterferon alpha-2b targets JAK2V617F-positive polycythemia vera cells in vitro and in vivo. Blood Cancer Journal. 2018;**8**(10):94. DOI: 10.1038/s41408-018-0133-0

[105] Greenfield G, McPherson S, Mills K, McMullin MF. The ruxolitinib effect: Understanding how molecular pathogenesis and epigenetic dysregulation impact therapeutic efficacy in myeloproliferative neoplasms. Journal of Translational Medicine. 2018;**16**(1):360. DOI: 10.1186/s12967-018-1729-7

[106] Kiladjian JJ, Guglielmelli P, Griesshammer M, Saydam G, Masszi T, Durrant S, et al. Efficacy and safety of ruxolitinib after and versus interferon use in the RESPONSE studies. Annals of Hematology. 2018;**97**(4):617-627. DOI: 10.1007/s00277-017-3225-1

[107] Bose P, Verstovsek S. Developmental therapeutics in myeloproliferative neoplasms. Clinical Lymphoma, Myeloma & Leukemia. 2017;**17S**:S43-S52. DOI: 10.1016/j.clml.2017.02.014

[108] Quintas-Cardama A, Vaddi K, Liu P, Manshouri T, Li J, Scherle PA, et al. Preclinical characterization of the selective JAK1/2 inhibitor INCB018424: therapeutic implications for the treatment of myeloproliferative neoplasms. Blood. 2010;**115**(15):3109-3117. DOI: 10.1182/blood-2009-04-214957

[109] Chase A, Bryant C, Score J, Haferlach C, Grossmann V, Schwaab J, et al. Ruxolitinib as potential targeted therapy for patients with JAK2 rearrangements. Haematologica. 2013;**98**(3):404-408. DOI: 10.3324/haematol.2012.067959

[110] Gozgit JM, Bebernitz G, Patil P, Ye M, Parmentier J, Wu J, et al. Effects of the JAK2 inhibitor, AZ960, on Pim/BAD/BCL-xL survival signaling in the human JAK2 V617F cell line SET-2. The Journal of Biological Chemistry. 2008;**283**(47):32334-32343. DOI: 10.1074/jbc.M803813200

[111] Harrison C, Vannucchi AM. Ruxolitinib: A potent and selective Janus kinase 1 and 2 inhibitor in patients with myelofibrosis. An update for clinicians. Therapeutic Advances in Hematology. 2012;**3**(6):341-354. DOI: 10.1177/2040620712459746

[112] Mazzacurati L, Lambert QT, Pradhan A, Griner LN, Huszar D, Reuther GW. The PIM inhibitor AZD1208 synergizes with ruxolitinib to induce apoptosis of ruxolitinib sensitive and resistant JAK2-V617F-driven cells and inhibit colony formation of primary MPN cells. Oncotarget. 2015;**6**(37):40141-40157. DOI: 10.18632/oncotarget.5653

[113] Spivak JL. Polycythemia vera: Myths, mechanisms, and management. Blood. 2002;**100**(13):4272-4290. DOI: 10.1182/blood-2001-12-0349

[114] Zeuner A, Pedini F, Francescangeli F, Signore M, Girelli G, Tafuri A, et al. Activity of the BH3 mimetic ABT-737 on polycythemia vera erythroid precursor cells. Blood. 2009;**113**(7):1522-1525. DOI: 10.1182/blood-2008-03-143321

[115] Will B, Siddiqi T, Jorda MA, Shimamura T, Luptakova K, Staber PB, et al. Apoptosis induced by JAK2 inhibition is mediated by Bim and enhanced by the BH3 mimetic ABT-737 in JAK2 mutant human erythroid cells. Blood. 2010;**115**(14):2901-2909. DOI: 10.1182/blood-2009-03-209544

[116] Winter PS, Sarosiek KA, Lin KH, Meggendorfer M, Schnittger S, Letai A, et al. RAS signaling promotes resistance to JAK inhibitors by suppressing BAD-mediated apoptosis. Science Signaling. 2014;**7**(357):ra122. DOI: 10.1126/scisignal.2005301

[117] Alonso-Alvarez S, Pardal E, Sanchez-Nieto D, Navarro M, Caballero MD, Mateos MV, et al. Plitidepsin: design, development, and potential place in therapy. Drug Design, Development and Therapy. 2017;**11**: 253-264. DOI: 10.2147/DDDT.S94165

[118] Garcia-Fernandez LF, Losada A, Alcaide V, Alvarez AM, Cuadrado A, Gonzalez L, et al. Aplidin induces the mitochondrial apoptotic pathway via oxidative stress-mediated JNK and p38 activation and protein kinase C delta. Oncogene. 2002;**21**(49):7533-7544. DOI: 10.1038/sj.onc.1205972

[119] Munoz-Alonso MJ, Gonzalez-Santiago L, Martinez T, Losada A, Galmarini CM, Munoz A. The mechanism of action of plitidepsin. Current Opinion in Investigational Drugs. 2009;**10**(6):536-542

[120] Gonzalez-Santiago L, Suarez Y, Zarich N, Munoz-Alonso MJ, Cuadrado A, Martinez T, et al. Aplidin induces JNK-dependent apoptosis in human breast cancer cells via alteration of glutathione homeostasis, Rac1 GTPase activation, and MKP-1 phosphatase downregulation. Cell Death and Differentiation. 2006;**13**(11):1968-1981. DOI: 10.1038/sj.cdd.4401898

[121] Losada A, Lopez-Oliva JM, Sanchez-Puelles JM, Garcia-Fernandez LF. Establishment and characterisation of a human carcinoma cell line with acquired resistance to Aplidin. British Journal of Cancer. 2004;**91**(7):1405-1413. DOI: 10.1038/sj.bjc.6602166

[122] Pardanani A, Tefferi A, Guglielmelli P, Bogani C, Bartalucci N, Rodriguez J, et al. Evaluation of plitidepsin in patients with primary myelofibrosis and post polycythemia vera/essential thrombocythemia myelofibrosis: results of preclinical studies and a phase II

clinical trial. Blood Cancer Journal. 2015;**5**:e286. DOI: 10.1038/bcj.2015.5

[123] Nguyen M, Cencic R, Ertel F, Bernier C, Pelletier J, Roulston A, et al. Obatoclax is a direct and potent antagonist of membrane-restricted Mcl-1 and is synthetic lethal with treatment that induces Bim. BMC Cancer. 2015;**15**:568. DOI: 10.1186/s12885-015-1582-5

[124] Hernandez-Ilizaliturri FJ, Khubchandani S, Olejniczak SH, Hosking P, Gruber E, Czuczman M. The BH3-mimetic obatoclax (GX15-070) posses a dual-mechanism of action and induces both apoptosis and autophagy-dependent cell death of B cell non-Hodgkin's. Lymphoma (B-NHL) Cells. Blood. 2008;**112**(11):605; 50th ASH Annual Meeting 2008; San Francisco; 6-9 December 2008

[125] Stamelos VA, Fisher N, Bamrah H, Voisey C, Price JC, Farrell WE, et al. The BH3 mimetic obatoclax accumulates in lysosomes and causes their alkalinization. PLoS One. 2016;**11**(3):e0150696. DOI: 10.1371/journal.pone.0150696

[126] Parikh SA, Kantarjian H, Schimmer A, Walsh W, Asatiani E, El-Shami K, et al. Phase II study of obatoclax mesylate (GX15-070), a small-molecule BCL-2 family antagonist, for patients with myelofibrosis. Clinical Lymphoma, Myeloma & Leukemia. 2010;**10**(4):285-289. DOI: 10.3816/CLML.2010.n.059

[127] Patel AB, Vellore NA, Deininger MW. New strategies in myeloproliferative neoplasms: The evolving genetic and therapeutic landscape. Clinical Cancer Research. 2016;**22**(5):1037-1047. DOI: 10.1158/1078-0432.CCR-15-0905

[128] Pemmaraju N, Carter BZ, Kantarjian HM, Cortes JE, Kadia TM, Garcia-Manero G, et al. Results for phase II clinical trial of LCL161, a SMAC mimetic, in Patients with Primary Myelofibrosis (PMF), Post-Polycythemia Vera Myelofibrosis (post-PV MF) or Post-Essential Thrombocytosis Myelofibrosis (post-ET MF). Blood. 2016;128(22):3105; 58th ASH Annual Meeting 2016; San Diego; 3-6 December 2016

Chapter 3

Endoplasmic Reticulum Stress-Mediated Cell Death

Mehtap Kara and Ezgi Oztas

Abstract

In normal functioning cells, endoplasmic reticulum (ER) is the major control site for folding, modification, and trafficking of secretory and cell-surface proteins. ER also plays a crucial role in the maintenance of cellular calcium homeostasis. Since ER is a key organelle in the cell; ER stress-mediated cell death can be associated with numerous diseases including Alzheimer disease, Parkinson disease, neuronal damage-induced ischemia, prion disease, cystic fibrosis, and diabetes mellitus. ER stress is a consequence of complex mechanisms which several cellular pathways interact with each other simultaneously. The two most important initiating points for ER stress-mediated cell death are; transcription factor CHOP/GADD153 and ER membrane protein kinase (IRE1). ER stress triggers proteolytic cleavage of caspase-12 and caspase-4, both of which are localized at the cytoplasmic side of the ER membrane to initiate the mechanism of cell death. Thus, ER stress and mitochondrial apoptosis are linked via caspase-12, which is seen in several degenerative diseases.

Keywords: endoplasmic reticulum stress, unfolded protein response signaling, autophagy, endoplasmic reticulum stress mediator proteins, endoplasmic reticulum stress-mediated diseases

1. Introduction

The endoplasmic reticulum (ER) is an intracellular organelle which has many roles in calcium storage, protein synthesis, degradation and transport, and carbohydrate and lipid metabolism. The ER has different types of domains in its specialized units to ensure its multifunction. The main function of the ER is the synthesis of secreted, cytosolic and membrane proteins. These processes are controlled by ribosomes that are localized in the cytosolic site of the ER. Initially, ribosomes and mRNA are united to form a translational complex on the inner surface of the cytosolic site of the ER. The protein synthesis starts in the cytosol and continues in mRNA-ribosome-signal recognition particle (SRP) complexes that are located on the ER membrane. Proteins classified simultaneously with protein synthesis are guided to the membrane or Golgi apparatus for secretion. The terminal step of the protein synthesis is the cleavage of the signal peptide; after this phase, proteins are secreted from the ER membrane to the cytosol via ribosomes [1, 2]. Between synthesis and secretion, exocrine proteins require folding and modifications through folding enzymes and chaperons. N-linked glycosylation, disulfide bond formation and oligomerization of proteins are determinants of the secretory proteins which indicates if they are or not [3, 4]. Hereby, the ER is one of the most crucial and multifunctional organelles for cell survival.

Since alterations of the ER's functions leads to unfolded and/or misfolded proteins in the cell, ER stress-mediated cell death underlie several serious diseases such as cardiovascular disease, neurodegeneration, ischemia and diabetes. Stress conditions are captured by transmembrane receptors which are localized on the ER and unfolded protein response (UPR) initiated by these receptors. Under chronic stress conditions, the adaptive response of the ER fails and the cells undergo mechanisms of cell death. In the ER stress conditions ATP, calcium and oxidizing environment are important factors for protein folding and disulfide bond formation. UPR is the major protective mechanism against deleterious and toxic effects of ER stress. Protein RNA-like ER kinase (PERK), activating transcription factor 6 (ATF6) and inositol-requiring enzyme 1 (IRE1α) are the tree main modulators of the ER stress response pathway [5, 6].

In a normal functioning cell PERK, ATF6 and IRE1 are in an inactive phase that is maintained by a specific ER chaperone, GRP78. When the ER stress is triggered, GRP78 is released to activate these three receptors as UPR. Unfolded protein stress has a crucial role in cell survival. However, the internal ribosomal entry site (IREs) bypasses the eukaryotic translation initiation factor 2α (eIF2α) controlling pathway. In the PERK-eIF2 pathway, activating transcription factor 4 (ATF4) is the key element which encodes cAMP response element-binding transcription factor (C/EBP) and promotes cell survival via modulation of redox reactions, stress response, protein synthesis and secretion. On the other hand ATF4 promotes C/EBP homologous protein (CHOP) which triggers apoptotic cell death. In the ER stress-mediated cell death, mitochondrial apoptotic pathway initiates the autophagy while other cell death mechanisms play a smaller role (**Figure 1**). It has been concluded that the ER stress-mediated cell death is associated with severe diseases including nervous system disorders, diabetes and cancer [7–10].

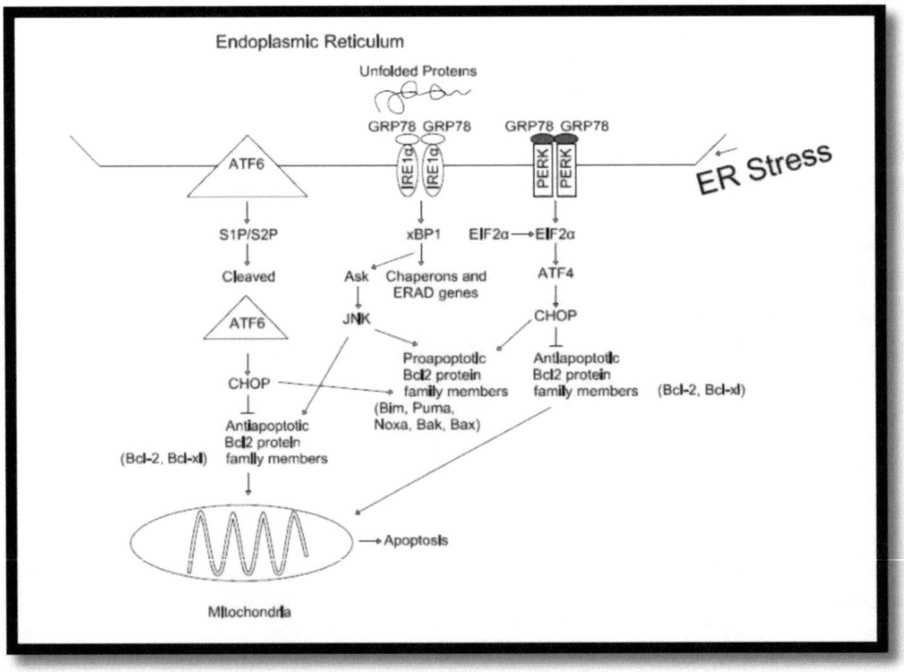

Figure 1.
ER stress-mediated pathways via PERK, IRE1α and ATF6 which stimulates apoptosis and suppress anti-apoptotic proteins.

2. Unfolded protein response (UPR) signaling

Several endogenous and exogenous factors may interfere with the ER protein folding mechanism and generate stress conditions which have been an issue of importance, strongly focused on in recent years. Chronic stress conditions may result with pathological perturbations in different systems in the organisms. Once the ER stress is triggered, UPR mechanisms strive to restore the ER homeostasis. If the UPR system does not be sufficient, apoptosis inducing signals are increased in the cell; and ultimately, cell death signaling pathways are activated [9, 10]. Enduring unfolded protein response (UPR) and ER stress in the cell cause dysfunctions of some mechanistic pathways which may in turn stimulate cell death. In the center of the UPR and ER stress response mechanism, IRE1α is placed as a key regulatory molecule. It has been demonstrated that IRE1α can directly bind to unfolded proteins and start signaling. IRE1α and its signaling pathway determine the fate of the cell between survival and death based on the longevity of ER stress [12].

PERK, ATF6 and IRE1 are the three main initiating proteins of UPR signaling due to ER stress. ATF6 is synthetized as an inactive precursor, and it contains bZIP transcription factor in its cytoplasmic domain. Under stress conditions, ATF6f, which is the active component of ATF6s, is released in the Golgi apparatus after cleavage by S1P and S2P proteases. ATF6f plays an important role as a transcription factor on ER homeostasis genes which include ER chaperons and ER-associated protein degradation (ERAD) [12–14]. PERK which is a transmembrane protein kinase gets in dimerization and auto-phosphorylation under ER stress conditions and phosphorylates eIF2α. Phosphorylated eIF2α have effects on initiating the selective translation of ATF4, protein folding factors genes expression regulation and plays a role in oxidative stress and amino acid metabolisms [14–16].

IRE1 is the most conserved signaling pathway in the ER stress mechanism. IRE1α and IRE1β are the main two isoforms of the IRE1. These isoforms have kinase and endoribonuclease activities at their cytoplasmic domain. Under stress conditions IRE1α goes into dimerization and auto-phosphorylation with a conformational change in its cytoplasmic part and activates the endoribonuclease domain. Active IRE1α catalyzes the splicing of X box-binding protein 1 (XBP-1) in its 26-nucleotide intron that result with active transcription factor XBP-1s which regulates protein folding, targeting to ER, ERAD and biogenesis of Golgi etc. [9, 16, 17]. IRE1α is a transmembrane protein that includes an N-terminal sensor domain, single transmembrane domain and C-terminal cytosolic effector domain. The C-terminal domain of IRE1α has both protein kinase and endoribonuclease activities. IRE1α oligomerization is induced by unfolded protein stress in the cell and following this cytosolic domain autophosphorylation occurs. With UPR control, IRE1α has an important cytoprotective effect [18, 19].

UPR signaling in the cell play an important role for restoring cellular homeostasis; however, chronic ER stress may result with cell death [11]. Apoptosis is the main cell death mechanism in ER stress; however, other types of cell death, such as necrosis, necroptosis or deregulated autophagy may contribute to ER stress too. Also autophagy that is a mechanism which enables the elimination of unfolded or misfolded proteins under ER stress conditions is one of the most studied issues in recent studies [20]. mRNA of IRE1α is regulated through X-box binding protein as well IRE1α controls its own mRNA expression by self-cleavage [21].

3. Cell death under endoplasmic reticulum stress conditions

3.1 PERK Signaling pathway

Under ER stress conditions, after activation of PERK, eIF2α phosphorilated by PERK and this phosphorylated eIF2α trigger translational arrest as a pro-survival response. This prosurvival response is important checkpoint step before cell death. Deficiencies of PERK expression or phosphorylation problems of eIF2α make cells more sensitive to ER stress conditions. PERK associated signaling pathway regulates important mechanisms such as autophagy, ATF4-mediated transcription pathway, and protein folding and redox metabolism [22, 23].

In different cell types it has been demonstrated that cell death is induced via the PERK signaling pathway under chronic stress conditions. The key molecule for initiating cell death is C/EBP homologous protein (CHOP), also named as growth arrest and DNA-damage-inducible 153 (GADD153). The expression of CHOPs is increased by ATF4. PERK activation induces eIF2α phosphorylation which in turn increases the selective transcription of ATF4 that increases CHOP level. Pro-apoptotic proteins such as GADD34, ERO1α (ER oxidase 1 alpha) and BH3-only proteins (BIM, PUMA and NOXA) expressions is increased by CHOP. PERK signaling pathway initiates mitochondrial apoptotic pathway. GADD34 and Ero1α increase cellular ROS production and calcium. Calcium release is regulated by IPR3 and the increase of cytosolic calcium triggers PTP related apoptosis in the cell. On the other hand, BIM, PUMA and NOXA induction cause cytochrome-c release from mitochondria via BAX and BAK activation [23–26].

CHOP, main member of Bcl-2 family, is one of the ER stress associated regulator molecule which downregulates Bcl-2. Another member of Bcl-2 family is BH3-only proteins which are pro-apoptotic proteins and upregulated by CHOP. Moreover, CHOP induces BIM, PUMA and NOXA expression levels [27–29]. CHOP has another important role in the ER as a modulator of oxidative status in the organelle. Increase in the level of CHOP and ERO1α in the cell causes a decrease of the glutathione (GSH) level which leads to ROS formation. ERO1α induces reduction of hydrogen peroxide (H_2O_2) in the ER lumen via reconstitution of the active state of proteins through re-oxidation of protein disulfide isomerases (PDIs). Increased ROS conditions in the cell via ER stress make cells sensitive to cell death. Moreover, CHOP increase in the cell triggers the activation of inositol- 1,4,5-trisphosphate receptor (IP3R) through ERO1α and calcium release from ER to cytosol which contribute to apoptosis. Thus, PERK plays a key role in inducing cell death via ROS production and calcium release. It has been demonstrated that, apoptosis can be induced without activation of the PERK signaling pathway; because, there are several triggers of apoptosis in different conditions. It has been reported in different studies that, PERK signaling deficiencies may play a role in different types of diseases such as Parkinson disease, diabetes, atherosclerosis, ALS, cardiac dysfunction and liver damage induced by alcohol. Further and detailed studies are needed to clarify the full mechanism of the ER stress dependent disease occurrence [24, 26, 30–35].

3.2 IRE1 signaling pathway

IRE1α/XBP-1 pathway plays a balancing role in survival-cell death homeostasis and also takes part in the gene regulation of the protein folding elements. With ER stress ASK1/JNK or NF-κB signaling pathways get activated by IRE1α-TRAF2 complex and afterwards cell death processes as apoptosis or autophagy starts in the cell [36–39]. IRE1α-dependent decay (RIDD) is IRE1α's endoribonuclease activity on several mRNAs. RIDD mechanism has a defensive role for degradation of proteins which have a misfolding potential and also RIDD takes part in pro-apoptotic mechanisms too. RIDD mechanism shows its pro-apoptotic effects through ER chaperons BiP/Grp78 mRNA degradation, effecting JNK signaling pathway or XBP-1 mRNA splicing. Thus, RIDD is placed in the center of the ER stress-mediated cell death and cell survival. It has been recently demonstrated that IRE1α show its endoribonuclease activity on different microRNAs (miRNA), caspase-2 and TXNIP which have role in cell death processes [40–42].

Dimerization, auto-phosphorylation and endoribonuclease domain engaging of IRE1α occurs under stress conditions. Active IRE1α activates transcription factor XBP1 which is named as XBP1s. XBP1 has a regulative role in protein folding [11, 12].

IRE1 contains a serine-threonine kinase and an endoribonuclease domain. With its endonuclease activity, IRE1 splices the 26-nucleotide intron from ATF6-induced XBP1 mRNA which generates the frameshift splice variant as sXBP1 and this variant encodes stable and active transcription factor. zXBP1 targets are ER chaperons and P58IPK which belongs to the HSP40 family. P58IPK plays a role in the negative feedback mechanism of PERK through binding and inhibiting PERK. P58IPK activity has the power to finish the UPR if the UPR could evade the ER stress, if not, the P58IPK activity gets suppressed and the apoptotic mechanism starts in the cell. Generally, IRE1 release has a strong pro-survival effect during stress conditions via UPR; however, long term active IRE induces kinase activities through the c-Jun N-terminal kinase (JNK) pathway and recruitment of TNF-receptor-associated factor 2 molecule (TRAF2). The IRE1-TRAF2 complex causes recruitment of apoptosis-signal-regulating kinase (ASK1) which in turn activates MAPKs JNK and p38. JNK activation associated with Bcl2 family members' regulation in different stress conditions. During ER stress JNK phosphorylates the Bcl2 and inhibits its anti-apoptotic function, however while JNK phosphorylates Bcl-2 homology domain 3 (BH3) and Bim, their pro-apoptotic features gets activated. As an important initiator of apoptosis, IRE1 is the last resorts for the ER stress regulated UPR after PERK and ATF6. IRE1 is the top step for modulation of pro-surviving or cell-death in the cell via ASK1 and JNK [5, 43].

3.3 ATF6 signaling pathway

GRP78 is one of the main regulatory proteins for the ER stress pathways, while ATF6 is separated from GRP78, ATF6 translocate to Golgi apparatus to get spliced from its active sites. Active ATF6 in turn enhance the gene expressions of stress response elements in the nucleus. GRP78, GRP94, protein disulfide isomerase, CHOP and XBP1 are some of the targets of ATF6. However, ATF6-mediated cell death mechanism has not been clarified yet and further studies are needed to explain detailed intracellular protein interactions [5].

ATF6 proteins ATF6α and ATF6β are regulatory proteins which belong to the bZIP transcription factor family. ATF6 binds to the ER membrane via its hydrophobic sequence. The ATF6 activation process during ER stress is different from the PERK and IRE1 activation processes. After GRP78 releases from ATF6, it translocates to the Golgi apparatus from the ER and the site1 and site2 proteases splice the ATF6s juxtamembrane site. After that, ATF6 translocates to the nucleus as a transcription factor for gene expression regulation. ATF6 stimulates homodimerization or heterodimerization of the ER stress related genes such as XBP1, IRE1, PDI, α-mannosidase-like protein 1 (EDEM1) as a result of misfolded protein degradation. In the literature it has been not clarified yet whether AFF6 regulates calcineurin 1 (RCAN1) which has calcium dependent pro-apoptotic functions [44, 45].

RCAN1's important substrates are pro-apoptotic Bcl-2 family members and Bcl-2 antagonist of cell death (BAD). Calcineurin dephosphorylates BAD and in turn

BAD dimerizes with the anti-apoptotic protein Bcl-Xl and inhibits its function. Cyclic AMP responsive element binding proteins such as CREB3l1 (oasis), CREB3l2, CREB3 (luman), CREB4, CREB-H are the other known ER stress transcription factors, however their mechanisms are not yet well detailed [46].

4. Endoplasmic reticulum stress mediator proteins

4.1 CHOP

C/EBP homologous protein (CHOP or DDIT3 or GADD153) is a bZIP transcription factor that regulates IRE1, PERK and ATF6 under ER stress conditions. ATF4, ATF6 and XBP1, important elements of UPR signaling, can bind to the CHOP gene promoter sequences to regulate its transcription [43, 44].

It has been demonstrated that with knock-out PERK and ATF4, CHOP induction was disrupted under ER stress, and also ATF2 and IRE1-ASK-p38 signaling pathways upregulated the activity of CHOP. CHOP induces apoptosis via inhibition of Bcl-2. CHOP and ERo1α together enhance the ER stress dependent protein loading in the cell which also enhances the apoptosis mechanism. Moreover, CHOP interacts with pro-apoptotic Bim to activate it and also inhibits Bcl-2, thus apoptosis occurs under ER stress condition. However, CHOP is not the main protein for ER stress-mediated cell death, it was demonstrated that ER stress-mediated apoptosis can occur without CHOP expression in $PERK^{-/-}$ and EIF2α (Ser51Ala) knock-in cells [27, 30, 46–48].

4.2 GADD34

GADD34 expression in the cell is associated with apoptotic cell death. As a protein phosphatase 1 (PP1)-interacting protein GADD34, dephosphorylates eIF2α which results with protein translation inhibition. The detailed mechanistic pathway which lays under GADD34-induced apoptosis has not been clarified yet. GADD34 expression may increase the pro-apoptotic proteins. It has been demonstrated in studies that blocking the GADD34 pathway in the cell can cause inhibition of R-stress-mediated apoptosis [5, 24].

4.3 BCL-2 proteins, calcium and caspases

Bcl2 family members are regulatory proteins of apoptosis which especially modulate the mitochondrial apoptotic pathway. In resting cells, on the mitochondria and ER membrane, the Bcl2 protein interacts with the pro-apoptotic proteins Bax and Bak and inhibits their functions. Moreover, dynein interacts with the pro-apoptotic protein Bim and inhibits its function. ER stress affects mitochondria and follows the same mechanism of the mitochondrial apoptotic pathway. During prolonged ER stress after activation of CHOP and JNK, Bim phosphorylates and releases from dynein. At the same time Bax and Bak unbound from Bcl2 and the execution phase of apoptosis initiates. During ER stress-mediated apoptosis cytochrome-c releases from mitochondria and apoptosome formation occurs [49, 50].

Detailed molecular information about the ER stress modulated apoptosis mechanism needs further investigations to clarify the different genes and proteins which play role in this mechanism, thus, new generation therapy models generation can be designed for ER stress-mediated apoptosis related diseases. It has been well known that CHOP and JNK play a central role in the activation of ER stress-mediated apoptosis. Bcl2 gene expression is inhibited by CHOP, which leads to the pro-apoptotic Bcl2 family members' activation. The functions of the Bcl2 family members also are regulated by JNK through phosphorylation. The IRE1-ASK1 pathway activates the JNK and JNK first phosphorylates the Bcl2 which is localized on the ER membrane. The BH3-only proteins on ER membrane gets activated and thus intracellular calcium flux become uncontrollable [51]. On the other hand JNK targets to BH3-only proteins are known as the "orchestrators of apoptosis." It has been reported that, the p53-upregulated modulator of apoptosis (PUMA), Noxa and Bim play roles in ER stress-mediated cell death. The Bim protein have three isoforms as short (BimS), and long (BimL and BimEL). Bim tethers dynein in the cell via its long isoforms BimL and BimEL under normal conditions, however under ER stress JNK phosphorylates Bim and stimulates its release from dynein leading to apoptosis initiation. Moreover, IRE1 induction directly activates Bax and Bak during ER stress-mediated apoptosis [52, 53].

All these activation processes are figured out with caspase activation. However, the caspase activation phase has not been defined clearly yet. It has been reported in different studies that caspases 12, 3, 6, 7, 8 and 9 do not play role in ER stress-mediated cell death. It has been speculated that caspase-4 maintains caspase-12's function in mammalians; however, it is not clarified yet [5]. It has been recently reported that in mammalians caspase-4, which is activated by ER stress, induces Bap31 and Bap20 proteins which play a role in the activation of the mitochondrial apoptotic pathway [54].

The Bcl2 family members (BCL-2 and BCL-XL) and transmembrane BAX inhibitor motif (TMBIM) (BI-1/TMBIM6 and GRINA/TMBIM3) interact and regulate IP3R activity. While cytosolic calcium level increases, BAX/BAX oligomerization (MOMP) occurs in the mitochondria and promotes the apoptotic pathway. ER foldase enzymes such as BiP and Protein disulfide isomerases (PDIs) translocate to the cytosol through BAX/BAX pores and in turn affect the plasma membrane pro-apoptotic protein Par-4 which results with apoptosis [11].

Caspase 12 and its polymorphic variant caspase 4 play an important promoting step role for ER stress-induced apoptosis. Caspase 12 activates pro-caspase-9, which belongs to the mitochondrial apoptotic pathway, without cytochrome-c release. Cytoplasmic calcium-activated protease calpain can activate caspase-12 by its cleavage and IRE1α also activates caspase-12 directly. Caspase-12 is located in the cell as a high molecular weight complex that includes apoptosis-linked gene-2 protein and p97 (ERAD mediator). P97 plays a crucial role for this pro-apoptotic stabilization of caspase-12. Once caspase-12 is activated, it activates the downstream caspases such as caspase-9 and caspase-7 which ultimately activates caspase-3; and hereby, apoptosis execution phase starts [55, 56].

5. Endoplasmic reticulum stress and autophagy

Endoplasmic reticulum (ER) stress is also closely related to the autophagy mechanism in the cell to maintain cellular homeostasis. Generation of autophagosomes occurs during the ER stress-induced autophagy process which encapsulates protein and damaged protein aggregates. As in the apoptosis process, PERK, IRE1 and ATF6 signaling pathways have role in initiating autophagy during ER stress, in addition Atg40/FAM134B takes part too [57]. Autophagy has crucial roles in maintaining optimum cellular activity, elimination of unfolded and misfolded proteins, elimination of defective organelles, protection against pathogens, balancing cellular energy storage, tumor suppression, biosynthesis of new molecules and cell death mechanisms [58].

ER stress UPR activating molecules PERK and IRE1α also activates the autophagy process by activating autophagy-related genes (Atg). Autophagosome formations generated by Atg proteins interact with the LC3II-PE complex. IRE1α-JNK pathway

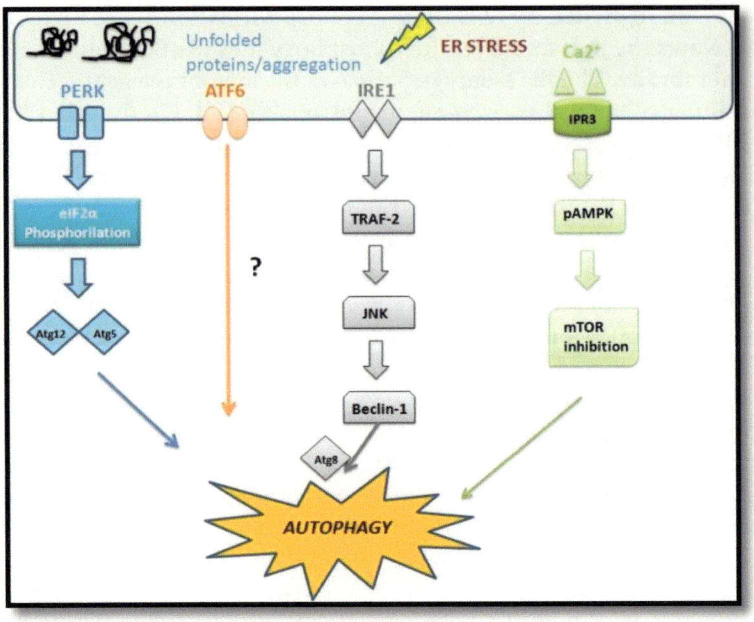

Figure 2.
ER stress-mediated autophagy induction via PERK, IRE1α, ATF6 and IPR3. PERK induce autophagy through eIF2α with Atg12 and Atg5 interaction, however ATF6 dependent pathway have not been clarified yet. Autophagy initiate by another pathway as Ca^{2+} influx through IPR3 transmembrane protein which inhibits mTOR. IRE1α pathway induce beclin-1 during autophagy induction processes.

activates Bcl2 by phosphorylation and subsequently stimulates Beclin1, ATG5 and ATG7. Thus, the Bcl2-Beclin complex dissociates and protein kinase-C activation phosphorylates LC3 and other autophagosome proteins. Another regulative process for autophagy is the mTOR pathway through AMP-dependent protein kinase (AMPK) that in turn induces the autophagy activating genes [59].

Autophagy process initiates EIF2AK3 activation which results with mTOR inhibition. Active EIF2AK3 upregulates ATF4 and subsequently SESN2, DDIT3 and DDIT4 are upregulated. These three active proteins inhibit the mTOR activity. AMPK pathway is activated via several types of metabolic stress, especially cellular energy state disorders and intracellular Ca level imbalance. AMPK pathway induces ULK1 and at the same time AMPK inhibits mTOR which has an inhibitory role on ULK1. MAPK8 and DAPK1 induction processes include formation of PtdIns3K and phosphorylation of the Bcl2-Beclin1 complex (**Figure 2**). All UPR sensory proteins have the potential to induce Beclin1 and Atg proteins for autophagy initiation. All UPR initiating molecules (IRE1, PERK, ATF4 and ATF6) activate the autophagy related protein Atg5-Atg12-Atg16 through CHOP activation. In mammalians, the ER stress-mediated autophagy mechanism is well defined and as an interesting point, under normal conditions the autophagy process plays a role for maintaining cellular homeostasis, however under stress conditions cellular mechanisms may inhibit the autophagy process with unknown regulative pathways. To understand the whole mechanism which takes part in ER stress-mediated autophagy, further studies are needed [57, 58].

6. Discussion

ER is pivotal organelle for cellular protein, lipid synthesis and Ca storage. During cell cycle process, ER is very active due to these physiological processes and addition due to its highly powerful stress response mechanism as UPR [60]. ER stress induces several different complex molecular pathways in the cell which may conclude physiological or pathological conditions. UPR signaling mechanism is one of the important cell protective homeostasis provider factor. UPR signaling rapidly initiate with IRE1α signaling after stress stimulus, secondly ATF6 pathway become a part of activity because of its slow kinetics and finally PERK mechanism step in. UPR mechanism have balancing role between cytoprotective and proapoptotic systems. Molecular features of ER stress and UPR mechanism is crucial for delivering targeted drugs for diseases which are associated with these signaling pathways [35]. There are several ongoing studies about ER stress mechanism to clarify signaling pathways, however many unknown mechanisms about pathway remain. As discussed in this chapter different signaling mechanisms play role in ER stress-mediated cell death however; pro-apoptotic mechanism components different from PERK, CHOP, Bcl-2 family members are not identified fully yet [11, 61, 62].

The ER stress mechanism plays an important role in several different diseases such as neurodegenerative diseases, ophthalmology disorders, inflammation diseases, viral infections, cancer, metabolic diseases, and atherosclerosis. It is important to understand the detailed mechanism which plays a role in ER stress-mediated diseases to provide more effective therapeutics. Alzheimer disease (AD), Parkinson's disease (PD), amyotrophic lateral sclerosis (ALS), prion diseases, retinitis pigmentosa, glaucoma, macular degeneration, as inflammatory bowel diseases, multiple sclerosis, rheumatoid arthritis, heart failure, cardiac hypertrophy, myocardial infraction and type I autoimmune diabetes are ER stress dependent diseases. In AD, mutant presenilin 1 interferes with the UPR mechanism and causes disruption in IRE1α, PERK and ATF6 signaling pathway and increase CHOP activity as a consequence of amyloid β-precursor protein (APP) accumulation in the neuron cells. Parkin is an important E3 ubiquitin ligase and it is also associated with ER stress-mediated cell death. The Mutant parkin gene causes accumulation of Lewy bodies in neurons that associated with defective UPR mechanism which result as Parkinson disease. ER stress mechanism depression is a very important strategy for cancer therapy. ER stress mechanism helps the tumor cells to adapt to its microenvironment. UPR plays a protective role for tumor cells and thus inhibition of ER stress could provide reducing in tumorigenesis. IRE1α/XBP1 has a crucial role in tumor angiogenesis. Considering all this together, ER stress and UPR pathways are important targets for chemotherapeutics [8, 44, 63].

7. Conclusion

ER stress-mediated cell death has a crucial role in several diseases pathophysiology. In recent years several studies have been done on the mechanism of ER stress, pathway details, its role in diseases and therapy. However, all these information are just the tips of the iceberg. To put forward effective therapeutic strategies, mechanistic pathway details should be defined well with further studies. ER stress seems to be a central mechanism to cell survival and cell death. Pathway associations with the other intracellular mechanisms are also needed to be clarified in order to understand the complexity.

Acknowledgements

The authors would like thank to Prof. Dr. Gül Özhan from Istanbul University, Faculty of Pharmacy, Department of Toxicology.

For valuable critiques and contributions, Sinem Beyaz for technical and graphical support and MSc. Ayşenur Günaydın and Pharm. Enes Bişirir for linguistic support.

This research received no specific grant from any funding agency in the public, commercial, private, or not-for-profit sectors.

Conflict of interest

The authors declare that there is no conflict of interest.

Author details

Mehtap Kara* and Ezgi Oztas
Department of Pharmaceutical Toxicology, Faculty of Pharmacy, Istanbul University, Istanbul, Turkey

*Address all correspondence to: mehtap.kara@istanbul.edu.tr

References

[1] Schwarz DS, Blower MD. The endoplasmic reticulum: Structure, function andresponse to cellular signaling. Cellular and Molecular Life Sciences. 2016;**73**(1):79-94

[2] Jan CH, Williams CC, Weissman JS. Principles of ER cotranslational translocation revealed by proximity-specific ribosome profiling. Science. 2014;**346**(6210):1257521

[3] Braakman I, Hebert DN. Protein folding in the endoplasmic reticulum. Cold Spring Harbor Perspectives in Biology. 2013;**5**(5):a013201

[4] van der Zand A, Gent J, Braakman I, Tabak HF. Biochemically distinct vesicles from the endoplasmic reticulum fuse to form peroxisomes. Cell. 2012;**149**(2):397-409

[5] Szegezdi E, Logue SE, Gorman AM, Samali A. Mediators of endoplasmic reticulum stress-induced apoptosis. EMBO Reports. 2006;**7**(9):880-885

[6] Doyle KM, Kennedy D, Gorman AM, Gupta S, Healy SJ, Samali A. Unfolded proteins and endoplasmic reticulum stress in neurodegenerative disorders. Journal of Cellular and Molecular Medicine. 2011;**15**(10):2025-2039

[7] Schröder M, Kaufman RJ. The mammalian unfolded protein response. Annual Review of Biochemistry. 2005;**74**:739-789

[8] Sano R, Reed JC. ER stress-induced cell death mechanisms. Biochimica et Biophysica Acta. 2013;**1833**(12):3460-3470

[9] Li J, Lee B, Lee AS. Endoplasmic reticulum stress-induced apoptosis: Multiple pathways and activation of p53-up-regulated modulator of apoptosis (PUMA) and NOXA by p53. The Journal of Biological Chemistry. 2006;**281**(11):7260-7270

[10] Chalmers F, van Lith M, Sweeney B, Cain K, Bulleid NJ. Inhibition of IRE1α-mediated XBP1 mRNA cleavage by XBP1 reveals a novel regulatory process during the unfolded protein response. Wellcome Open Research. 2017;**2**:36

[11] Urra H, Dufey E, Lisbona F, Rojas-Rivera D, Hetz C. When ER stress reaches a dead end. Biochimica et Biophysica Acta. 2013;**1833**(12):3507-3517

[12] Gardner BM, Walter P. Unfolded proteins are IRE1-activating ligands that directly induce the unfolded protein response. Science. 2011;**333**(6051):1891-1894

[13] Haze K, Yoshida H, Yanagi H, Yura T, Mori K. Mammalian transcription factor ATF6 is synthesized as a transmembrane protein and activated by proteolysis in response to endoplasmic reticulum stress. Molecular Biology of the Cell. 1999;**10**(11):3787-3799

[14] Asada R, Kanemoto S, Kondo S, Saito A, Imaizumi K. The signalling from endoplasmic reticulum-resident bZIP transcription factors involved in diverse cellular physiology. Journal of Biochemistry. 2011;**149**(5):507-518

[15] Vattem KM, Wek RC. Reinitiation involving upstream ORFs regulates ATF4 mRNA translation in mammalian cells. Proceedings of the National Academy of Sciences. 2004;**101**(31):11269-11274

[16] Harding HP, Zhang Y, Zeng H, Novoa I, Lu PD, Calfon M, et al. An integrated stress response regulates amino acid metabolism and resistance to oxidative stress. Molecular Cell. 2003;**11**(3):619-633

[17] Calfon M, Zeng H, Urano F, Till JH, Hubbard SR, Harding HP,

et al. IRE1 couples endoplasmic reticulum load to secretory capacity by processing the XBP-1 mRNA. Nature. 2002;**415**(6867):92-96

[18] Lee AH, Iwakoshi NN, Glimcher LH. XBP-1 regulates a subset of endoplasmic reticulum resident chaperone genes in the unfolded protein response. Molecular and Cellular Biology. 2003;**23**(21):7448-7459

[19] Shamu CE, Walter P. Oligomerization and phosphorylation of the Ire1p kinase during intracellular signaling from the endoplasmic reticulum to the nucleus. The EMBO Journal. 1996;**15**(12):3028-3039

[20] Ogata M, Hino S, Saito A, Morikawa K, Kondo S, Kanemoto S, et al. Autophagy is activated for cell survival after endoplasmic reticulum stress. Molecular and Cellular Biology. 2006;**26**:9220-9231

[21] Hassler J, Cao SS, Kaufman RJ. IRE1, a double-edged sword in pre-miRNA slicing and cell death. Developmental Cell. 2012;**23**:921-923

[22] Koumenis C, Naczki C, Koritzinsky M, Rastani S, Diehl A, Sonenberg N, et al. Regulation of protein synthesis by hypoxia via activation of the endoplasmic reticulum kinase PERK and phosphorylation of the translation initiation factor eIF2alpha. Molecular and Cellular Biology. 2002;**22**:7405-7416

[23] Lin JH, Li H, Zhang Y, Ron D, Walter P. Divergent effects of PERK and IRE1 signaling on cell viability. PLoS One. 2009;**4**:e4170

[24] Marciniak SJ, Yun CY, Oyadomari S, Novoa I, Zhang Y, Jungreis R, et al. CHOP induces death by promoting protein synthesis and oxidation in the stressed endoplasmic reticulum. Genes and Development. 2004;**18**:3066-3077

[25] Kojima E, Takeuchi A, Haneda M, Yagi A, Hasegawa T, Yamaki K, et al. The function of GADD34 is a recovery from a shutoff of protein synthesis induced by ER stress: Elucidation by GADD34-deficient mice. The FASEB Journal. 2003;**17**:1573-1575

[26] Li G, Mongillo M, Chin KT, Harding H, Ron D, Marks AR, et al. Role of ERO1-alpha-mediated stimulation of inositol 1,4,5-triphosphate receptor activity in endoplasmic reticulum stress-induced apoptosis. Journal of Cell Biology. 2009;**186**(6):783-792

[27] Puthalakath H, O'Reilly LA, Gunn P, Lee L, Kelly PN, Huntington ND, et al. ER stress triggers apoptosis by activating BH3-only protein Bim. Cell. 2007;**129**:1337-1349

[28] Reimertz C, Kogel D, Rami A, Chittenden T, Prehn JH. Gene expression during ER stress-induced apoptosis in neurons: Induction of the BH3-only protein Bbc3/PUMA and activation of the mitochondrial apoptosis pathway. The Journal of Cell Biology. 2003;**162**:587-597

[29] Cazanave SC, Elmi NA, Akazawa Y, Bronk SF, Mott JL, Gores GJ. CHOP and AP-1 cooperatively mediate PUMA expression during lipoapoptosis. American Journal of Physiology. Gastrointestinal and Liver Physiology. 2010;**299**:G236-G243

[30] McCullough KD, Martindale JL, Klotz LO, Aw TY, Holbrook NJ. Gadd153 sensitizes cells to endoplasmic reticulum stress by down-regulating Bcl2 and perturbing the cellular redox state. Molecular and Cellular Biology. 2001;**21**:1249-1259

[31] Tavender TJ, Bulleid NJ. Molecular mechanisms regulating oxidative activity of the Ero1 family in the endoplasmic reticulum. Antioxidants and Redox Signaling. 2010;**13**:1177-1187

[32] Malhotra JD, Kaufman RJ. Endoplasmic reticulum stress and oxidative stress: A vicious cycle or a double-edged sword? Antioxidants and Redox Signaling. 2007;**9**:2277-2293

[33] Verfaillie T, Rubio N, Garg AD, Bultynck G, Rizzuto R, Decuypere JP, et al. PERK is required at the ER–mitochondrial contact sites to convey apoptosis after ROS-based ER stress. Cell Death and Differentiation. 2012;**19**:1880-1891

[34] Silva RM, Ries V, Oo TF, Yarygina O, Jackson-Lewis V, Ryu EJ, et al. CHOP/GADD153 is a mediator of apoptotic death in substantia nigra dopamine neurons in an in vivo neurotoxin model of parkinsonism. Journal of Neurochemistry. 2005;**95**:974-986

[35] Lin JH, Li H, Yasumura D, Cohen HR, Zhang C, Panning B, et al. IRE1 signaling affects cell fate during the unfolded protein response. Science. 2007;**318**(5852):944-949

[36] Han D, Lerner AG, VandeWalle L, Upton JP, Xu W, Hagen A, et al. IRE1alpha kinase activation modes control alternate endoribonuclease outputs to determine divergent cell fates. Cell. 2009;**138**:562-575

[37] Nishitoh H, Matsuzawa A, Tobiume K, Saegusa K, Takeda K, Inoue K, et al. ASK1 is essential for endoplasmic reticulum stress-induced neuronal cell death triggered by expanded polyglutamine repeats. Genes and Development. 2002;**16**:1345-1355

[38] Castillo K, Rojas-Rivera D, Lisbona F, Caballero B, Nassif M, Court FA, et al. BAX inhibitor-1 regulates autophagy by controlling the IRE1alpha branch of the unfolded protein response. The EMBO Journal. 2011;**30**:4465-4478

[39] Hu P, Han Z, Couvillon AD, Kaufman RJ, Exton JH. Autocrine tumour necrosis factor alpha links endoplasmic reticulum stress to the membrane death receptor pathway through IRE1alpha-mediated NF-kappaB activation and down-regulation of TRAF2 expression. Molecular and Cellular Biology. 2006;**26**:3071-3084

[40] Hollien J, Lin JH, Li H, Stevens N, Walter P, Weissman JS. Regulated IRE1-dependent decay of messenger RNAs in mammalian cells. Journal of Cell Biology. 2009;**186**:323-331

[41] Bouchecareilh M, Higa A, Fribourg S, Moenner M, Chevet E. Peptides derived from the bifunctional kinase/RNase enzyme IRE1alpha modulate IRE1alpha activity and protect cells from endoplasmic reticulum stress. The FASEB Journal. 2011;**25**:3115-3129

[42] Upton JP, Wang L, Han D, Wang ES, Huskey NE, Lim L, et al. IRE1alpha cleaves select microRNAs during ER stress to derepress translation of proapoptotic caspase-2. Science. 2012;**338**:818-822

[43] Yoshida H, Matsui T, Yamamoto A, Okada T, Mori K. XBP1 mRNA is induced by ATF6 and spliced by IRE1 in response to ER stress to produce a highly active transcription factor. Cell. 2001;**107**(7):881-891

[44] Samali A, Fitzgerald U, Deegan S, Gupta S. Methods for monitoring endoplasmic reticulum stress and the unfolded protein response. International Journal of Cell Biology. 2010;**2010**:830307

[45] Bravo R, Parra V, Gatica D, Rodriguez AE, Torrealba N, Paredes F, et al. Endoplasmic reticulum and the unfolded protein response: Dynamics and metabolic integration. International Review of Cell and Molecular Biology. 2013;**301**:215-290

[46] Kim I, Xu W, Reed JC. Cell death and endoplasmic reticulum stress:

Disease relevance and therapeutic opportunities. Nature Reviews. Drug Discovery. 2008;7(12):1013-1030

[47] Ma Y, Brewer JW, Diehl JA, Hendershot LM. Two distinct stress signaling pathways converge upon the CHOP promoter during the mammalian unfolded protein response. Journal of Molecular Biology. 2002;**318**:1351-1365

[48] Oyadomari S, Mori M. Roles of CHOP/GADD153 in endoplasmic reticulum stress. Cell Death and Differentiation. 2004;**11**:381-389

[49] Boya P, Cohen I, Zamzami N, Vieira HL, Kroemer G. Endoplasmic reticulum stress-induced cell death requires mitochondrial membrane permeabilization. Cell Death and Differentiation. 2002;**9**:465-467

[50] Zong WX, Li C, Hatzivassiliou G, Lindsten T, Yu QC, Yuan J, et al. Bax and Bak can localize to the endoplasmic reticulum to initiate apoptosis. Journal of Cell Biology. 2003;**162**:59-69

[51] Bassik MC, Scorrano L, Oakes SA, Pozzan T, Korsmeyer SJ. Phosphorylation of BCL-2 regulates ER Ca^{2+} homeostasis and apoptosis. The EMBO Journal. 2004;**23**:1207-1216

[52] Shamas-Din A, Brahmbhatt H, Leber B, Andrews DW. BH3-only proteins: Orchestrators of apoptosis. Biochimica et Biophysica Acta. 2011;**1813**(4):508-520

[53] Iurlaro R, Muñoz-Pinedo C. Cell death induced by endoplasmic reticulum stress. The FEBS Journal. 2016;**283**(14):2640-2652

[54] Rosati E, Sabatini R, Rampino G, De Falco F, Di Ianni M, Falzetti F, et al. Novel targets for endoplasmic reticulum stress-induced apoptosis in B-CLL. Blood. 2010;**116**(15):2713-2723

[55] Morishima N, Nakanishi K, Takenouchi H, Shibata T, Yasuhiko Y. An endoplasmic reticulum stress-specific caspase cascade in apoptosis. Cytochrome c-independent activation of caspase-9 by caspase-12. The Journal of Biological Chemistry. 2002;**277**(37):34287-34294

[56] Boyce M, Yuan J. Cellular response to endoplasmic reticulum stress: A matter of life or death. Cell Death and Differentiation. 2006;**13**:363-373

[57] Song S, Tan J, Miao Y, Zhang Q. Crosstalk of ER stress-mediated autophagy and ER-phagy: Involvement of UPR and the core autophagy machinery. Journal of Cellular Physiology. 2018;**233**(5):3867-3874

[58] Rashid HO, Yadav RK, Kim HR, Chae HJ. ER stress: Autophagy induction, inhibition and selection. Autophagy. 2015;**11**(11):1956-1977

[59] Singh MP, Kang SC. Endoplasmic reticulum stress-mediated autophagy activation attenuatesfumonisin B1 induced hepatotoxicity in vitro and in vivo. Food and Chemical Toxicology. 2017;**110**:371-382

[60] Schwarz DS, Blower MD. The endoplasmic reticulum: Structure, function and response to cellular signaling. Cellular and Molecular Life Sciences. 2016;**73**(1):79-94

[61] Lin JH, Li H, Yasumura D, Cohen HR, Zhang C, Panning B, Shokat KM, Lavail MM, Walter P. IRE1 signaling affects cell fate during the unfolded protein response. Science. 9 Nov 2007;**318**(5852):944-949

[62] Harding HP, Zhang Y, Bertolotti A, Zeng H, Ron D. Perk is essential for translational regulation and cell survival during the unfolded protein response. Molecular Cell. May 2000;**5**(5):897-904

[63] Lin JH, Walter P, Yen TS. Endoplasmic reticulum stress in disease pathogenesis. Annual Review of Pathology. 2008;**3**:399-425

Chapter 4

Cell Death Mechanisms of the Promising Anticancer Compound Gallotannin

Marwa Houssein and Hala Gali-Muhtasib

Abstract

The polyphenolic hydrolyzable tannin, gallotannin (GT), also known as tannic acid, possesses interesting anticarcinogenic properties. An evidence from experimental studies suggests that GT is effective against multiple cancer types. Gallotannin has been shown to induce programmed cell death in a wide variety of cancers including colon, breast, prostate, and liver, among others. Apoptosis, cellular senescence, autophagy, and necrosis are the main mechanisms by which GT can suppress cancer progression. In addition, GT is a potent inhibitor of many proliferation pathways. Herein, this chapter provides a summary of our current knowledge about GT's programmed cell death mechanisms against cancer.

Keywords: programmed cell death, apoptosis, autophagy, necroptosis, cellular senescence, polyphenols, gallotannin, cancer

1. Introduction

Over the past decade, cancer was considered as the major public health concern and one of the leading causes of death worldwide with 9.6 million cancer deaths reported in 2018. The main recurrent cancers are lung, colorectal, stomach, liver, prostate, and breast. Various external (tobacco, radiation, infections, nutrition) and internal (mutations, hormones, and immune conditions) factors contribute to cancer incidence. Described as a group of diseases characterized by uncontrolled proliferation and growth of abnormal cells, cancer is mainly characterized by its resistance to cell death and sustained activation of proliferation pathways. Throughout life, there is a balance between cell death and cell proliferation [1]. Any imbalance favoring one process over the other would contribute to the development of significant disorders such as autoimmune diseases, cancer, AIDS, neurodegenerative disorders, myocardial infarctions, atherosclerosis, and insulin-dependent diabetes. Furthermore, cell death can be divided into two forms: regulated and accidental. Unlike regulated cell death which relies on the activation of signal transduction cascades and can be modulated pharmacologically or genetically, accidental cell death occurs immediately in response to physical (high pressure, osmotic forces, temperatures, etc.), chemical (extreme pH variations, etc.), or mechanical (shear forces, etc.) cues that occur in an uncontrollable form [2]. Besides apoptosis, there are many forms of programmed cell death, namely, necroptosis, cellular senescence, autophagy, slow cell death, and paraptosis [3].

IntechOpen

One third of all cancers can be prevented by a healthy lifestyle which includes an appropriate and balanced nutrition [4]. There is a growing interest in using dietary compounds as preventive and therapeutic agents in cancer due to their relative safety, their immediate action on target tissues, and their specific action against cancer cells [5]. The protective effects of dietary compounds are a consequence of different modes of action such as their ability to neutralize carcinogens, hamper the transcription of oncogenes, activate detoxifying enzymes, and trigger cell death in mutated and cancerous cells.

Polyphenols are a group of phytochemicals that when consumed decrease the risk of chronic diseases especially cancer. They are effective in treating solid tumors by inducing a cohort of effects such as cell cycle arrest and cell death [6]. The polyphenolic hydrolyzable tannin, gallotannin, and penta-1, 2, 3, 4, 6-O-galloyl--beta-D-glucose (PGG), a precursor of GT, have been shown to exert various biological effects ranging from anti-inflammatory to anticancer effects in various tumor cells [7]. Thus, current research aims at developing therapeutic approaches that would benefit from the interconnected matrix of signaling pathways and the ability of natural compounds to induce programmed cell death processes that could prevent and restrain tumor development.

2. Programmed cell death mechanisms

2.1 Apoptosis

Each day, our body eliminates, via apoptosis, billions of unwanted cells. In addition to being essential for development and homeostasis, apoptosis is the major mechanism of programmed cell death. During apoptosis, apoptotic cells undergo characteristic changes in cell morphology, including shrinkage of nuclei, nuclear chromatin condensation, cytoplasmic shrinkage, dilated endoplasmic reticulum, membrane blebbing, and the exposure of specific phagocytic signaling molecules on the cell surface [8]. The contents of the cell are englobed in apoptotic bodies which are then recognized by the phagocytic cells and digested in lysosomes. In most cells, apoptosis leads to the activation of caspases which mediates the auto destruction of the cell. Caspases, a family of cysteine proteases, exist as inactive precursors (pro-caspases) and are activated upon cleavage. The C-terminal side of a four amino acid motif, X-X-X-Asp (where X can be any amino acid) is the preferred cleavage site for the known caspases. Activated caspases in turn cleave various intracellular and cytoplasmic membrane substrates, leading to cellular disintegration [9].

2.1.1 Apoptotic pathways

In the mammalian system, two major pathways lead to apoptosis: the intrinsic pathway which involves the mitochondria and the extrinsic pathway which is initiated by death receptors [10]. The intrinsic pathway of apoptosis is mediated by the Bcl-2 family protein (also known as mitochondrial or stress pathway). Bcl-2 family, composed of both anti-apoptotic and pro-apoptotic proteins, is generally divided into three subgroups based on their roles in apoptosis and the BH regions they share: one anti-apoptotic group and two pro-apoptotic groups. The anti-apoptotic group includes Bcl-2, Bcl-xL, Bcl-w, Bcl-B, A1, and Mcl-1, which share three or four BH regions. The pro-apoptotic Bcl-2 family members include Bax, Bak, Bcl-xs, Bok, and Bcl-GL, which have two or three BH domains. Another pro-apoptotic group contains the BH3-only proteins, including Bad, Bid, Bim, Bik, Noxa, Puma, Bcl-Gs, Blk, Bmf, and Hrk, which share only the BH3 domain. A key event of the intrinsic pathway is the mitochondrial outer membrane permeabilization (MOMP) process, which is considered the point-of-no return in apoptosis induction. Normally, the anti-apoptotic members of the Bcl2 family prevent MOMP [11]. In response to stress stimuli such as oncogenes, direct DNA damage, oxidative stress, and starvation, two pro-apoptotic proteins, Bax and Bak, become activated by BH3-only proteins that serve as sensors for apoptotic stimuli. Once activated, Bax and Bak permeabilize the outer membrane of mitochondria, causing the release of pro-apoptotic factors such as cytochrome c. In the cytosol, cytochrome c binds to monomeric apoptotic protease activating factor-1 (APAF-1) at its WD40 domain and induces a conformational change in APAF-1 promoting APAF-1 oligomerization and initiating the formation of the apoptosome. APAF-1 then binds to pro-caspase 9 resulting in its auto-cleavage and release of active caspase 9. Active caspase 9 then cleaves the effector caspases, such as caspase 3 and caspase 7, resulting in their activation and promoting the cell death process. The extrinsic pathway or death receptor pathway involves the binding of ligands to cell surface "death receptors" which in turn initiate the caspase cascade. Death receptors, located on the cell membrane, are members of the TNFR family and are characterized by the presence of a death domain (DD) that plays a crucial role in apoptotic signal transduction [12]. The best characterized ligands and their corresponding death receptors include TNF-α/TNF-R1, FasL/FasR, APO3L/DR3, TRAIL/TRAIL-R1, TRAIL/TRAIL-R2, and TRADD/DR6. The binding of death ligands results in the oligomerization and the activation of the death receptors. Oligomerization of the receptors is followed by binding of specific adapter proteins (FADD, TRADD) to their receptor, which in turn leads to the activation of the caspase signaling pathway. FADD binds to pro-caspase 8 through its dead effector domain (DED) allowing the formation of DISC, the death-inducing signaling complex, and the autocatalytic activation of pro-caspase-8. Active caspase-8 executes the apoptotic process through direct cleavage and activation of effector caspases (caspases 3, 6, and 7) [13].

2.1.2 Apoptosis and cancer

The evasion of apoptosis is one of the prominent hallmarks of cancer cells. This is the result of mutations in apoptosis-related genes. The Bcl-2 family members, Fas, p53, and c-Myc are the common mutated genes in cancer. Overall, malignant cells, in different kinds of cancers, have an anti-apoptotic phenotype with low level of pro-apoptotic proteins such as Bax and high level of anti-apoptotic proteins such as Bcl-2 and Bcl-xL. The tumor suppressor p53 is mutated in most cancers including colorectal carcinoma, brain and lung cancer, mammary carcinoma, and skin and bladder carcinomas. The overexpression of inhibitor apoptosis proteins (IAP) and downregulation of surface death receptors (CD95, DR4, and DR5) were also detected in cancer cells [14].

2.2 Cellular senescence

Senescence is a normal process caused by telomere shortening after successive cell divisions of normal somatic cells. This process irreversibly halts the cell from proliferation. In addition to being a normal process, senescence can be activated in response to oncogenic activation, oxidative stress, and DNA damage. Morphologically, senescent cells can be distinguished by their enlarged, flattened, and granular morphology; nuclear enlargement; and altered chromatin structure [15]. Biochemically, cellular senescence is characterized by an enhanced β-galactosidase activity, inhibition of cyclin-dependent kinases (CDKs), the absence of proliferation markers, and the presence of senescence-associated heterochromatin foci (SAHF) [2].

2.2.1 Pathways of cellular senescence

The molecular pathways of senescent cells are not unique but differ between cells from different species and among different cell types from the same species. The heterogeneous pathways of senescence, however, meet at p53 and p-Rb [16]. In the p53-p21 pathway, p53 is the important player of the senescence response. The expression of the phosphorylated form of p53 increases in senescent cells leading to an increase in its transcriptional activity. P53 is activated in response to shortened telomeres that activate a DNA damage cascade through ATM/ATR and Chk1/Chk2 resulting in G1 phase arrest. P53 is also activated in response to DNA damage, oxidative stress, and activation of *Ras* oncogene leading to telomere-independent premature senescence. One of the most important targets of p53 is p21 which can also be expressed in a p53-independent manner [17]. Some cell types have been found to undergo senescence upon overexpressing p21 and escape senescence upon its deletion. Mouse embryonic fibroblasts lacking p21 undergo senescence suggesting that senescence can occur in a p21-independent manner in these rodent cells [18]. In the p16-pRb pathway, p53 and p-Rb are thought to act simultaneously to achieve senescence because their concomitant inactivation is needed to terminate the senescence response in human cells. The phosphorylation state of Rb controls the progression through the cell cycle. When phosphorylated by cyclin-dependent kinases, p-Rb liberates E2F transcription factor that transcribes target genes responsible for DNA replication and progression through the cell cycle. Inhibition of CDKs by cyclin-dependent kinase inhibitors prevents the phosphorylation of Rb. When hypophosphorylated, Rb binds E2F, thus preventing the transcription of E2F target genes [19]. CDKIs belong to two families: the CIP/KIP family including p21, p27, and, p57 and the INK4 family including p15, p16, p18, and p19. P16 expression is found to be high in senescent cells driven to cell cycle arrest by stressful stimuli such as DNA damage, oxidative stress, and oncogenic Ras activation. Both p53-p21 and p16-pRb senescent pathways converge on inhibiting the phosphorylation of Rb [20]. In some cases, senescence is hindered by the inhibition of either p53 or p-Rb which suggests a sole p53 and p-Rb signaling pathway. In other cases, inhibiting senescence requires the inactivation of both p53 and p-Rb supporting the existence of two simultaneous pathways. This variation in the molecular mechanisms of the senescence signaling pathway depends on several factors including p16 expression, the type of cell line, culture conditions, and the amount of stress [21].

2.2.2 Cellular senescence and cancer

Senescence represents a fail-safe mechanism that guards against oncogenic transformation [22]. Cancer arises after a normal cell accumulates several mutations that are inherited with each replicative cycle. Senescence helps limit the accumulation of mutations by restricting the replicative ability of normal cells, thereby preventing cancer development [23]. Furthermore, it has been shown that initial oncogenic events will lead to senescence, and at that point the senescence-associated secretory phenotype will induce immune clearance limiting early tumor growth [24]. But many cancerous cells acquire indefinite proliferation by escaping senescence through different mechanisms that inhibit telomere shortening such as triggering telomerases or increasing telomere length by homologous recombination or inhibiting tumor suppressors. In fact, various tumor suppressors and oncogenes have been shown to regulate senescence in normal cells, and senescence bypass appears to be an important step in the development of cancer [25]. For instance, inhibiting BRAF-induced senescence by the loss of the tumor suppressor PTEN will lead to melanoma progression [18]. Thus, evading senescence is an important step toward full malignancy and metastasis. Tumor cells may escape senescence indirectly by acquiring mutations that affect senescence-related proteins such as the tumor suppressor proteins p53 and p-Rb which are frequently mutated in cancer [26]. Furthermore, many studies detected senescent cells in premalignant mice and human tumors and not in the malignant tumors. This senescence response was thought to have an antitumorigenic role in cancer-predisposed tissues [27].

Although most cancer cells bypass oncogene-induced senescence, various studies described the anticancer ability of chemotherapeutic drugs to induce senescence in those cancer cells, an event which was termed as therapy-induced senescence.

2.3 Autophagy

In response to starvation, hypoxic conditions and high temperatures, and DNA and organelle damage, the autophagy cell death process is activated. Double-membrane cytoplasmic vesicles called autophagosomes engulf cytoplasmic organelles and fuse to lysosomes to form autolysosomes when the cellular components are digested [28].

2.3.1 Pathways of autophagy

Autophagy is driven by Atgs proteins and is regulated by PI3 kinase types I and III. PI3K type 1 inhibits autophagy through PDK1 and AKT which regulate mTOR. Atg6 (Beclin1) is part of PI3K type 3 complex which promotes the nucleation of autophagic vesicles. In resting conditions, mTOR phosphorylates and inactivates ULK1 (Atg1). Metabolic stress activates AMPK which inhibits mTOR activity and activates ULK1 leading for the activation of the autophagy function of PI3K type 3 through Beclin1 phosphorylation [29]. The proteins Atg1 and Atg13 allow the membrane isolation. Elongation is then mediated by Atg10, Atg5, and ATG12. The recruitment of the cytoplasmic protein LC3 by Atg7, Atg4, and Atg3 to the nascent autophagosome is necessary for the expansion and fusion events. The formation of autolysosomes is not only sufficient for autophagic cell death; additional death signals are still needed. Enhanced expression of c-Jun N-terminal kinase (JNK) generates such signals and rapidly induces autophagy cell death [30].

2.3.2 Autophagy and cancer

Decreased rate of autophagic activity is related to tumorigenesis, and autophagic cell death does not occur in most cancer cells. Beclin1 is downregulated in many cancer types including prostate, breast, and ovarian cancer [31]. Mutations of Atg5 and LC3 (microtubule-associated protein 1 light chain 3B) promote myeloma and glioblastoma, respectively [32]. JNK activation is significantly decreased in cancer cells [33].

2.4 Necrosis

Although most scientists considered necrosis as accidental cell death and an uncontrolled process, several studies showed that some forms of necrosis are programmed [34]. Necroptosis is controlled by receptor interacting protein (RIP) kinases, NADPH oxidases, poly(ADP-ribose) polymerase-1 (PARP1), and calpains [35]. During necroptosis, the disruption of the cellular membrane integrity leads to the release of the intracellular materials in the extracellular environment which induce a local inflammatory response. These perturbations are detected by specific death receptors such as FAS and TNFR1 or pathogen recognition receptors (PRRs) such as TLR3 and TLR4 [2]. The activation of receptors allows the activation of RIPK1 which in turn activates RIPK3. The signaling complex formed by RIPK1 and RIPK3 is known as necrosome and leads to the phosphorylation of mixed lineage kinase domain-like pseudokinase (MLKL). The activation of MLKL results in the formation of MLKL oligomers that upon translocation to the plasma membrane trigger its permeabilization [30].

2.4.1 Necroptosis and cancer

Cancer cells evade necroptosis by downregulation or functional mutations of RIP1, RIP3, and MLKL genes [36]. Colon cancer cells showed decreased levels of RIP1 and RIP3 expression [37]. A hypermethylation in the promoter region of RIP3 was detected in lung cancer cells resulting in the loss of RIP3 expression [38]. Mutations in RIP1, RIP3, and MLKL were observed in human cancer tissues which modulate the RIP kinase interaction with other proteins or decrease their activity [39].

3. Polyphenols

Higher plants are found to have thousands of molecules with polyphenol structures, and edible plants are also found to contain several hundreds of these polyphenolic molecules. Plants generate these molecules as secondary metabolites that have crucial roles in plant development, reproduction, and pigmentation. They also have other roles like providing protection against diseases, parasites, ultraviolet radiation, and predators [40]. Collectively, these chemicals are known as phytochemicals which are categorized into five groups: carotenoids, phenolics, alkaloids, nitrogen-containing compounds, and organosulfur compounds. Tannin, a subgroup of phenolics and to which GT belongs, is further categorized into two biologically and chemically distinct subtypes, the condensed and hydrolyzable tannins [41]. Condensed tannins, depending on their size, could be either soluble or insoluble, whereby the oligomers are soluble, and the polymers are not. On the other hand, the hydrolyzable tannins, also referred to as galloyl and hexahydroxydiphenol esters, are composed of a β-D-glucose unit, linked to at least five galloyl groups through ester bonds. GT, or tannic acid, belongs to this last group that is also characterized by the presence of digalloyl residues consisting of meta-depside bonds between two galloyl groups [42].

3.1 Polyphenols and cancer

Plant polyphenols were found to suppress tumor invasion and metastasis, such is the case of green tea polyphenols, hydrolyzable tannins, grape seed, curcumin, and resveratrol [43]. This inhibition potential is due to the downregulation of the matrix metalloproteases (MMPs) by the natural phenolics [44]. Also, epigallocatechin gallate (EGCG), the most abundant polyphenol found in tea, has been reported to decrease the vascular endothelial growth factor (VEGF), thus inhibiting tumor angiogenesis [45]. Indeed, there has been increased evidence over the years of the effect of polyphenols on the fate of cancer cells leading to growth, differentiation, and apoptosis. Phytochemicals contribute to cancer prevention by interfering in the different stages of cancer development from tumor initiation and throughout all the hallmarks of cancer [4]. Thus, one of the most studied and acclaimed biological effects of polyphenols relates to its antioxidant properties; indeed polyphenols are able to scavenge reactive oxygen species (ROS) including radical and nonradical oxygen species such as O_2^-, HO^-, NO^-, H_2O_2, O_2, and HOCl, as well as oxidatively generated free radicals ROC and ROOC such as those derived from biomolecules such as low-density lipoproteins (LDLs), proteins, and oligonucleic acids (DNA and RNA) [46]. Different human intervention studies on the health potential of polyphenols were conducted in healthy volunteers or on high-risk developing cancer individuals. A one-dose diet rich in polyphenols, such as fruit juices, chocolate, strawberries, and grape seed concentrate, was able to reduce the antioxidant status and protect from oxidative stress [39]. In another study, hemodialysis patients consumed 200 mL/day of red fruit juice, and the results showed a significant decrease in DNA oxidation damage and NF-κB binding activity. In summary polyphenols exert their preventive effects on cancer cells via different mechanisms, mainly via their antioxidant effects, antiproliferation and antisurvival effects, induction of cell cycle arrest, induction of apoptosis, anti-inflammatory effects, and inhibition of metastasis and angiogenesis [47].

4. Gallotannin

Gallotannin or tannic acid is a hydrolyzable tannin that is characterized by the formation of meta-depside bonds between two galloyl groups resulting in digalloyl residues [42].

4.1 Bioavailability, biodegradation, and absorption

Gallotannin is mainly present in mangoes, pomegranate, acorns, walnuts, and beverages such as wine [48]. The metabolism of tannic acid was studied in a rat model system [49]. Based on this study, the authors proposed that administered tannic acid was not hydrolyzed by the acidity of the stomach but rather by tannase enzymes produced by intestinal bacteria. Upon hydrolysis, tannic acid released glucose and small phenolic acids which include gallic acid (GA), 4-O-methylgallic acid (4-OMGA), resorcinol (RE), pyrogallol (PY), and ellagic acid (EA) [50]. It has been postulated that GT hydrolysis could be measured by the release of gallic acid. Gallic acid is further metabolized by the colon microflora into pyrogallol which is ready to be absorbed by the intestinal cells [51].

4.2 Biosafety and toxicity

Dietary polyphenols are considered as safe and well-tolerated compounds. Only few studies have reported toxicity and adverse effects induced by plant polyphenols [52]. Tested on human epithelial and fibroblast cells, PGG did not show any toxicity at concentrations below 50 μM [53]. Galla Rhois is the excrescence formed by parasitic aphids, primarily *Schlechtendalia chinensis* Bell, on the leaf of sumac. The gallotannin-enriched extract isolated from Galla Rhois (GEGR) was administered orally to ICR mice using three different concentrations (250, 500, and 1000 mg/kg body weight) for 14 days to evaluate its hepatotoxicity and nephrotoxicity. No toxicity on the liver and kidney organs was detected in the mice when GT was used at less than 1000 mg/kg [54]. The Chinese medicinal plant rich in PGG, *Galla chinensis*, was also studied for its acute and subchronic oral toxicity. No acute oral toxicity was produced in rats using the oral dose of 5760 mg/kg. The no-observed--adverse-effect level was lesser than 1500 mg/kg body weight/day in the subchronic oral toxicity study. This dose is three times higher than the recommended dose for clinical application. In addition, the *Galla chinensis* serum did not show any side effects to rats in the central nervous system, cardiovascular system, and respiratory system [55]. Although these findings consider that GT and PGG are nontoxic compounds, further investigations are needed to prove their biosafety.

4.3 Functional properties

Regarding its pharmacological potential, GT has been suggested to prevent several diseases and possess different activities [56]. Gallotannin inhibited microbial growth of both Gram-positive and Gram-negative pathogens through its antimicrobial activity [57]. Gallotannin precursor, PGG, had an antiviral activity versus Herpes simplex virus type 1 (HSV-1) [58] and a high anti-inflammatory effect against atherosclerosis [59]. This effect was mediated by poly(ADP-ribose) glycohydrolase (PARG) pathway, whereby GT was found to inhibit PARG and thus trigger nuclear accumulation of poly (ADP-ribose) (PAR). Poly ADP-ribosylation is a posttranslational modification of proteins operated by poly (ADP-ribose) polymerases (PARPs). Studies have shown that inhibition of PAR formation impairs the expression of several genes involved in the inflammatory response. The antioxidant activity of PGG was associated with the inhibition of prooxidant enzymes [60]. The antidiabetic effect of PGG was mediated by the activation of insulin receptor associated with the transportation of glucose in the adipocytes and the reduction of blood glucose and insulin level in diabetic animals [61]. The anticancer effect of GT will be discussed in detail in the next section. **Figure 1** summarizes the mechanisms of action of GT on cancer cells.

4.4 Mechanisms of anticancer activity of gallotannin

Gallotannin is a polyphenol that possesses interesting anticarcinogenic properties. The effect of GT in different cancer cell lines was explored by many researchers. Different programmed death mechanisms by GT were detected in different kinds of cancers and even in different cell lines of one type of cancer. **Table 1** represents GT anticancer effect.

4.4.1 Gallotannin and apoptosis

Gallotannin has the ability to induce apoptosis in HCT116 (p53−/−) and HCT116 (p21−/−) colon cancer cell lines through the induction of Bax/Bcl-1

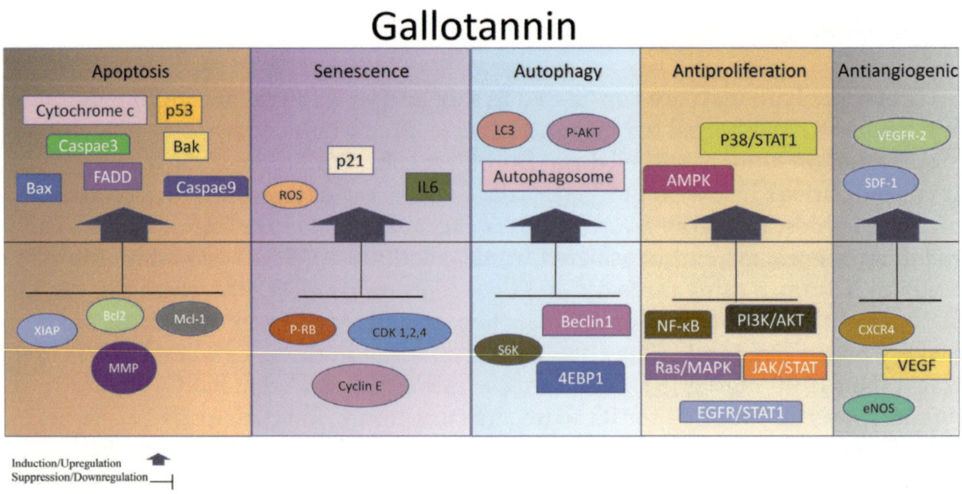

Figure 1.
Schematic diagram of the anticancer mechanisms of Gallotannin. GT induces apoptosis, senescence and autophagy and inhibits cellular proliferation and angiogenesis.

Cancer type	Cell lines	Cell death type	Molecular mechanism	Reference
Colon	HCT116 $p53^{-/-}$ HCT116 $p21^{-/-}$	Apoptosis	Increased Bax/Bcl2 ratio	[65]
	HCT116 $p53^{-/-}$ HCT116 $p21^{-/-}$ HCT116 $p53^{+/+}$	Senescence	Enhanced ROS	[64]
Liver	HepG2	Apoptosis	Pro-caspase3/9; Bcl2; PARP cleavage	[68]
	HepG2 SK-Hep1	Senescence	SA-β-gal activity; p21; IL6	[71] [81]
	SK-Hep1 HepG2	Autophagy	LC3; LC3B-II; Beclin1	[81]
Lung	A549	Apoptosis	Caspase3; p53	[69]
	A549	Senescence	IL6	[81]
	A549	Autophagy	LC3-II	
Prostate	DU145 PC-3, m2182	Apoptosis	Caspases; Mcl-1 UPR pathway	[70] [72]
	TRAMP-C2 PC-3	Autophagy	AKT; S6K; 4EBP1	[84]
Leukemia T-cells	Jurkat	Apoptosis	Bax	[73]
Breast	MDA-MB-231 MCF-7	Apoptosis	Caspase3 Bak; FADD	[74] [75]
	MCF-7 SKBr3	Senescence	IL6 ROS	[81] [82]
	MCF-7	Autophagy	LC3-II	[81]
Cervical	HeLa	Apoptosis	Loss of MMP	[76]
		Necrosis	Increase sub-G1 cells	
Leukemia	HL-60RG	Apoptosis	ROS	[77]
Esophageal	TE-2	Apoptosis	Bax; Bcl2; XIAP	[62]
Gingival	GSCC	Apoptosis	Bcl2; Bax; cytochrome c	[63]
Acute myeloid leukemia	HL-60	Apoptosis	Caspases; PARP cleavage; cytochrome c	[78]
Glioma (brain)	HS 683	Apoptosis	Caspases3/9; PARP; ROS	[79]

Table 1.
Anticancer effects of GT and its programmed cell death mechanisms and molecular targets.

protein level [65]. Gold nanoparticle formulated tannic acid (AuNP-TA) was also used against colorectal cancer, and apoptosis was detected in HCT116 cell line through the upregulation of the expression of caspase3 and 9, Bak, and Bax; loss of mitochondrial membrane potential; and release of cytosolic cytochrome c [66]. In liver cancer HepG2 hepatocellular carcinoma cell line, GT attenuated the expression of pro-caspase9, pro-caspase3, Bcl2, and integrin β1 and cleaved poly(ADP)-ribose polymerase [68]. Gallotannin dramatically induced apoptosis through the expression of p53 and active caspase-3 and fragmented DNA in A549 human lung carcinoma cells [69]. In DU145, PC-3, and M2182 prostate cancer cell lines, the inhibition of Mcl-1 and activation of caspases were critically involved in GT-induced apoptosis [70]. Another study using prostate cancer cell lines showed that tannic acid can promote apoptosis via the ER stress-mediated UPR pathway [72]. The inhibition of the proteasome by TA in Jurkat T cells was associated with Kip1 accumulation of the cyclin-dependent kinase inhibitor p27 and pro-apoptotic protein Bax and was accompanied by the induction of G_1 arrest and apoptosis [73]. Gallotannin regulates apoptosis via increased expression of active caspase-3 and cyclooxygenase-2 (COX-2) expression through PI3-kinase and p38 kinase pathway in MDA-MB-231 human breast cancer cells [74]. In the human breast adenocarcinoma MCF-7 cells, GT was able to cause apoptosis by increasing the percentage ratios of apoptotic proteins Bak and FADD [75]. Gallic acid induced apoptosis and/or necrosis in cervical cancer HeLa cell line, which was accompanied by the loss of mitochondrial membrane potential (MMP; DeltaPsi(m)) [76]. Intracellular ROS induced by gallic acid, especially H2O2, plays an important role in eliciting an early signal in apoptosis in leukemia HL-60RG cells [77]. Gallic acid induced apoptosis in esophageal cancer cells (TE-2) via the upregulation of Bax and downregulation of anti-apoptosis proteins such as Bcl2 and XIAP [62]. Tannic acid induced apoptosis in gingival squamous cell carcinoma (GSCC) via the inhibition of Bcl-2 and increase of the mitochondrial localization of Bax leading to the loss of mitochondrial membrane potential, resulting in the release of cytochrome c to the cytosol [63]. The combination of tannic acid with As_2O_3 induced apoptosis in acute myeloid leukemia (AML) HL-60 cell line through the activation of the caspase cascade, cleavage of poly (ADP-ribose) polymerase, disruption of mitochondrial membrane potential, and release of cytochrome c [78]. Treated with GT, HS 683, a glioma cell line, showed an activation of pro-caspase 3 and caspase 9, cleavage of poly (ADP-ribose) polymerase, loss of mitochondrial membrane potential, and increased intracellular ROS production [79].

4.4.2 Gallotannin and cellular senescence

In HCT116 human colon cancer cells wildtype for p53 and p21 and null for these genes, GT caused senescence independent of p21 and p53 with the partial involvement of ROS in this senescence effect [64]. Gallotannin increased the subG1 population and induced senescence via upregulation of p21 and caused G1 arrest and higher SA-β-gal activity in hepatocellular carcinoma (HCC) HepG2 and SK-Hep1 cell lines [71]. The GT precursor (PGG) was found to induce S-phase and G1 cell cycle arrest in breast [80] and prostate cancer cells [67]. HepG2 human liver cancer cells, MCF-7 human breast cancer cells, and A549 human lung cancer underwent senescence upon treatment with PGG. These cells acquired enlarged and flattened morphology which was associated with a significant increase in the level of both *IL6* mRNA and its secretory protein which is the key component of senescence-associated secretory phenotype (SASP) [81]. Senescence like S response, also known as premature senescence, is mediated by the intracellular ROS generation. PGG induced senescence like S phase arrest in HepG2, Huh-7 human hepatoma cells and SKBr3 human breast cancer [82].

4.4.3 Gallotannin and autophagy

In SK-Hep1 hepatocellular carcinoma cell line, GT induced autophagic features by increasing LC3 punctate, LC3B-II conversion, autophagic vacuoles, and decreased expression of Beclin1 [71]. HepG2 human liver cancer cells, MCF-7 human breast cancer cells, and A549 human lung cancer cells treated with PGG increased the LC3-II level, a sign of autophagy [81]. A new molecular nanoparticle based on iron(III)-tannic complexes (Fe–TA NPs) induced autophagic cell death in HepG2 cells via upregulation of LC3 mRNA expression and autophagosome formation [83]. Autophagic responses were observed in human DU145 and PC-3 prostate cancer xenografts in nude mice detected by the formation of autophagosomes. As for molecular changes, a rapid inhibition of the phosphorylation of mammalian target of rapamycin-downstream targets S6K and 4EBP1 was observed by PGG in PC-3 and TRAMP-C2 cells but not that of mammalian target of rapamycin itself, along with increased AKT phosphorylation [84].

4.4.4 Gallotannin and necrosis

Ammar et al. showed a necrotic effect of the crude extract of *Terminalia chebula retz.*, fruit containing tannic acid. Tested on many cancer cell lines derived from many cancer types such as breast, prostate, and osteosarcoma, the extract induced necrosis when used at higher concentrations [85]. A similar effect was detected when GT was used at higher doses in HeLa cells whereby cell death was marked by an increase of sub-G1 cells [76].

4.4.5 Antiangiogenic effect of gallotannin

Angiogenesis is defined as a process by which new blood vessels are formed and is a crucial process for tumor progression. The activation of VEGFR-2 is specific for vascular endothelial cells to promote migration during angiogenesis. Gallotannin effectively inhibited this process through downregulation of VEGF leading to the inhibition of endothelial cell angiogenesis. The inhibition of VEGF was correlated with the reduction of eNOS, a mediator of vasodilatation [86]. An upregulation of SDF-1 and inhibition of CXCR4 allowed the mobilization of pro-angiogenic hematopoietic cells. Gallotannin dramatically decreased the SDF-1/CXCR4 interaction which suppressed tumor angiogenesis [87].

4.4.6 Antiproliferation effects of gallotannin

Gallotannin was found to interfere with the activation of the proliferation pathways including NF-κB, PI3K/AKT/mTOR, JAK/STAT, and many others. In colon cancer, GT inhibited NF-κB pathway [88]. In breast cancer, GT caused ROS-dependent upregulation of AMPK and downregulation of the AKT/mTOR pathway [89] and PI3K/AKT pathway [90], while it inhibited EGFR/STAT1/3 and enhanced p38/STAT1 signaling pathway [91]. Gallotannin also inhibited JAK2/STAT3 pathway in gingival squamous cell carcinoma [63]. Ras/MAPK is also suppressed by GT [87]. Gallotannin was found to inhibit JAK/STAT pathway in HCT116 colon cancer cells by our group (Gali-Muhtasib et al., unpublished findings).

5. Conclusion

With its pleiotropic molecular mechanisms of action on cells ranging from apoptosis, senescence, autophagy, necrosis, antiproliferative, and antiangiogenic effects, Gallotannin seems to be an interesting compound from natural sources with minimal side effects and high effectivity against different types of cancer. Further research including preclinical and clinical studies may propose the use of gallotannin as an anticarcinogenic drug.

Conflict of interest

The authors declare no conflict of interest.

Author details

Marwa Houssein[1,2] and Hala Gali-Muhtasib[2,3]*

1 Department of Biological Sciences, Faculty of Science, Beirut Arab University, Lebanon

2 Center for Drug Discovery, American University of Beirut, Lebanon

3 Department of Biology and Department of Anatomy, Cell Biology and Physiological Sciences, American University of Beirut, Lebanon

*Address all correspondence to: amro@aub.edu.lb

References

[1] Czabotar PE, Lessence G, Strasser A, Adams JM. Control of apoptosis by the BCL-2 protein family: Implications for physiology and therapy. Nature Reviews Molecular Cell Biology. 2014;**15**:49-63. DOI: 10.1038/nrm3722

[2] Lorenzo G, Ilio V, Guido K. Molecular mechanisms of cell death: Recommendations of the Nomenclature Committee on cell death 2018. Cell Death and Differentiation. 2018;**25**: 486-541. DOI: 10.1038/s41418-017-0012-4

[3] Broker LE, Kruyt FA, Giaccone G. Cell death independent of caspases: A review. Clinical Cancer Research. 2005; **11**:3155-3162. DOI: 10.1158/1078-0432.CCr-04-2223

[4] Gonzalez-Vallinas M, Gonzalez-Castejon M, Rodriguez-Casado A. Dietary phytochemicals in cancer prevention and therapy: A complementary approach with promising perspectives. Nutrition Reviews. 2013;**71**:585-599. DOI: 10.111/nure.12051

[5] Spatafora C, Tringali C. Natural-derived polyphenols as potential anticancer agents. Anti-Cancer Agents in Medicinal Chemistry. 2012; **12**:902-918. DOI: 10.2174/187152012802649996

[6] Lucci P, Saurina J, Núñez O. Trends in LC-MS and LC-HRMS analysis and characterization of polyphenols in food. TrAC, Trends in Analytical Chemistry. 2016;**88**:1-24. DOI: 10.1016/j.trac.2016.12.006

[7] Shaikh Q, Yang M, Memon K, Lateef M, Na W, Jiang T. 1,2,3,4,6-Pentakis[-O-(3,4,5-trihydroxybenzoyl)]-a, b-D-glucopyranose (PGG) analogs: Design, synthesis, anti-tumor and anti-oxidant activities. Carbohydrate Research. 2016; **430**:72-81. DOI: 10.1016/jcarres.2016.04.021

[8] Hotchkiss RS, Strasser A, McDunn JE, Swanson PE. Cell death. The New England Journal of Medicine. 2009;**361**:1570-1583. DOI: 10.1056/NEJMRA0901217

[9] Guangwu X, Yufang S. Apoptosis signaling pathways and lymphocyte homeostasis. Cell Research. 2007;**17**: 759-771. DOI: 10.1038/CR.2007.52

[10] Adams JM. Ways of dying: Multiple pathways to apoptosis. Genes & Development. 2003;**17**:2481-2495. DOI: 10.1101/gad.1126903

[11] Fulda S, Debatin K-M. Extrinsic versus intrinsic apoptosis pathways in anticancer chemotherapy. Oncogene. 2006;**25**:4798-4811. DOI: 10.1038/sj.onc.1209608

[12] Schleich K, Krammer PH, Lavrik IN. The chains of death: A new view on caspase-8 activation at the DISC. Cell Cycle. 2013;**12**:193-194. DOI: 10.4161/cc.23464

[13] Wilson NS, Dixit V, Ashkenazi A. Death receptor signal transducers: Nodes of coordination in immune signaling networks. Nature Immunology. 2009;**10**: 348-355. DOI: 10.1038/ni.1714

[14] Fan X, Ya J. Apoptosis in oncology. Cell Research. 2001;**11**:1-7. DOI: 10.1038/sj.cr.7290060

[15] Tchkonia T, Zhu Y, Deursen JV. Cellular senescence and the senescent secretory phenotype: Therapeutic opportunities. The Journal of Clinical Investigation. 2013;**123**(3):966-972. DOI: 10.1172/JCI64098

[16] Norman E, Ronald A. Crime and punishment. Nature. 2005;**436**:636-637. DOI: 10.1038/436636a

[17] Aravinthan A. Cellular senescence: A hitchhiker's guide. Human Cell. 2015; **28**(2):51-64. DOI: 10.1007/s13577-015-0110-x

[18] Rufini A, Tucci P, Celardo I, Melino G. Senescence and aging: The critical roles of p53. Oncogene. 2013;**32**: 5129-5143. DOI: 10.1038/onc.2012.640

[19] Sabatino M, Petiti J, Sosa L. Evidence of cellular senescence during the development of estrogen-induced pituitary tumors. Endocrine-Related Cancer. 2015;**22**:299-317. DOI: 10.1530/ERC-14-0333

[20] Mowla S, Lam E, Jat P. Cellular senescence and aging: The role of B-MYB. Aging Cell. 2014;**13**:773-779. DOI: 10.1111/acel.12242

[21] Falandry C, Bonnefoy M, Freyer G. Biology of cancer and aging: A complex association with cellular senescence. Journal of the American Society of Clinical Oncology. 2014;**32**:2604-2610. DOI: 10.1200/JCO.2014.55.1432

[22] Bertolotto C. Metabolism under the spotlight in senescence. Pigment Cell & Melanoma Research. 2014;**27**:3-5. DOI: 10.1111/pcmr.12184

[23] Lee M, Lee J. Exploiting tumor cell senescence in anticancer therapy. BMB Reports. 2014;**47**:51-59. DOI: 10.5483/BMBREP.2014.47.2.005

[24] Coates PJ. In brief: Cell senescence. The Journal of Pathology. 2013;**230**: 239-240. DOI: 10.1002/path.4182

[25] Campisi J. Aging, cellular senescence, and cancer. Annual Review of Physiology. 2013;**75**:685-705. DOI: 10.1146/annurev-physiol-030212-183653

[26] He L, Ying C, Liling T. Cellular senescence regulated by SWI/SNF complex subunits through p53/p21 and p16/pRB pathway. The International Journal of Biochemistry & Cell Biology. 2017;**90**:29-37. DOI: 10.1016/j.biocel.2017.07.007

[27] Collado M, Serrano M. Senescence in tumours: Evidence from mice and humans. Nature Reviews. Cancer. 2010; **10**:51-57. DOI: 10.1038/nrc2772

[28] Tsujimoto Y, Shimizu S. Another way to die: Autophagic programmed cell death. Cell Death and Differentiation. 2005;**12**:1528-1534. DOI: 10.1038/sj.cdd.4401777

[29] Bialik S, Santosh K, Kimchi A. Autophagy-dependent cell death–Where, how and why a cell eats itself to death. Journal of Cell Science. 2018;**131**: jcs215152. DOI: 10.1242/jcs.215152

[30] Shimizu S, Konishi A, Nishida Y, Mizuta T, Nishina H, Yamamoto A, et al. Involvement of JNK in the regulation of autophagic cell death. Oncogene. 2010;**29**:2070-2082. DOI: 10.1038/onc.2009.487

[31] Qu X, Yu J, Bhagat G, Furuya N, Hibshoosh H, Troxel A, et al. Promotion of tumorigenesis by heterozygous disruption of the beclin 1 autophagy gene. The Journal of Clinical Investigation. 2003;**112**:1809-1820. DOI: 10.1172/JCI20039

[32] Huang X, Bai HM, Chen L, Li B, Lu YC. Reduced expression of LC3B-II and Beclin 1 in glioblastoma multiforme indicates a down-regulated autophagic capacity that relates to the progression of astrocytic tumors. Journal of Clinical Neuroscience. 2010;**17**:1515-1519. DOI: 10.1016/j.jocn.2010.03.051

[33] Shigeomi S, Tatsushi Y, Masatsune T, Satoko A. Autophagic cell death and cancer. International Journal of Molecular Sciences. 2014;**5**:3145-3153. DOI: 10.3390/ijms15023145

[34] Marcus P. Apoptosis meets necrosis. Nature. 2011;**471**:310-312. DOI: 10.1038/47310a

[35] Ouyang L, Shi Z, Zhao S, Wang F-T, Bao J-K. Programmed cell death pathway sin cancer: A review of apoptosis, autophagy and programmed necrosis. Cell Proliferation. 2012;**45**: 487-498. DOI: 10.1111/J.1365-2184. 2012.00845.x

[36] Su Z, Yang Z, Xie L, Dewitt JP, Chen Y. Cancer therapy in the necroptosis era. Cell Death and Differentiation. 2016;**23**:748-756. DOI: 10.1038/cdd.2016.8

[37] Moriwaki K, Bertin J, Gough P, Orlowski G, Chan FK. Differential roles of RIPK1 and RIPK3 in TNF-induced necroptosis and chemotherapeutic agent-induced cell death. Cell Death & Disease. 2015;**6**:e1636. DOI: 10.1038/ cddis.2015.16

[38] Fukasawa M, Kimura M, Morita S, Matsubara K, Yamanaka S, Endo C, et al. Microarray analysis of promoter methylation in lung cancers. Journal of Human Genetics. 2006;**51**:368-374. DOI: 10.1007/S10038-005-0355-4

[39] Forbes S, Bhamra G, Bamford S, Dawson E, Kok C, Clements J, et al. The catalogue of somatic mutations in cancer (COSMIC). Current Protocols in Human Genetic. 2008; Chapter 10: unit 10.11. DOI: 10.1002/0471142905.hg 1011s57

[40] Araujo JR, Goncalves P, Martel F. Chemopreventive effect of dietary polyphenols in colorectal cancer cell lines. Nutrition Research (New York, N.Y.). 2011;**3**:77-87. DOI: 10.1016/j. nutres.2011.01.006

[41] Gali-Muhtasib H, Perchellet J, Khatib S. Inhibitory effects of plant tannins on ultraviolet light-induced epidermal DNA synthesis in hairless mice. Photochemistry and Photobiology. 1998;**67**:663-668. DOI: 10.1111/j.1751-1097.1998.tb0911.x

[42] Gross G. From lignins to tannins: Forty years of enzyme studies on the biosynthesis of phenolic compounds. Phytochemistry. 2008;**69**:3018-3031. DOI: 10.1016/j.phytochem.2007.04.031

[43] Dai J, MumpeR R. Plant phenolics: Extraction, analysis and their antioxidant and anticancer properties. Molecules. 2010;**15**:7313-7352. DOI: 10.3390/molecules15107313

[44] Tanimura S, Kadomoto R, Tanaka T. Suppression of tumor cell invasiveness by hydrolyzable tannins (plant polyphenols) via the inhibition of matrix metalloproteinase-2/-9 activity. Biochemical and Biophysical Research Communications. 2005;**330**: 1306-1313. DOI: 10.1016/J.BBRC. 2005.03.116

[45] Ramos S. Cancer chemoprevention and chemotherapy: Dietary polyphenols and signalling pathways. Molecular Nutrition & Food Research. 2008;**52**: 507-526. DOI: 10.1002/mnfr.200700326

[46] Quideau S, Deffieux D, Douat-Casassus C, Pouysegu L. Plant polyphenols: Chemical properties, biological activities, and synthesis. Angewandte Chemie (International Ed. in English). 2011;**50**:586-621. DOI: 10.1002/anie.201000044

[47] Hassan R, Mohammad H, Reza K. Polyphenols and their benefits: A review. International Journal Of food Properties. 2017;**20**:1700-1704. DOI: 10.1080/10942912.2017.1354017

[48] Beecher GR. Overview of dietary flavonoids: Nomenclature, occurrence and intake. The Journal of Nutrition. 2003:3248S-3254S. DOI: 10.1093/jn/ 133.10.3248S

[49] Nakamura Y, Tsuji S, Tonogai Y. Method for analysis of tannic acid and its metabolites in biological samples: Application to tannic acid metabolism in the rat. Journal of Agricultural and Food Chemistry. 2003;**51**:331-339. DOI: 10.1021/jf020847

[50] Singh B, Bhat T, Sharma O. Biodegradation of tannic acid in an in vitro ruminal system. Livestock Production Science. 2001;**68**:259-262. DOI: 10.1016/S0301-6226(00)002727-X

[51] Revero I, Rodriguez H, Campos G, Curiel JA, Ascaso C. Tannic acid-dependent modulation of selected lactobacillus plantarum traits linked to gastrointestinal survival. PLoS One. 2013;**8**(6):e66473. DOI: 10.1371/journal.pone.0066473

[52] Thomasset SC, Berry DP, Garcea G, Marczylo T, Steward WP, Gescher AJ. Dietary polyphenolic phytochemicals-promising cancer chemopreventive agents in humans? A review of their clinical properties. International Journal of Cancer. 2007;**120**(3):451-458. DOI: 10.1002/ijc.22419

[53] Lin M, Chang F, Hua M, Wu Y, Liu S. Inhibitory effects of 1,2,3,4,6-penta-O-galloyl-beta-D-glucopyranose on biofilm formation by Staphylococcus aureus. Antimicrobial Agents and Chemotherapy. 2011;**55**(3): 1021-1027. DOI: 10.1128/AAC.00843-10

[54] Jun G, Ji-Eun K, Eun-Kyoung K, Sung-Hwa S, Ji-Eun S, Chan-Kyu P, et al. Hepatotoxicity and nephrotoxicity of gallotannin-enriched extract isolated from Galla Rhois in ICR mice. Laboratory Animal Research. 2015; **31**(3):101-110. DOI: 10.5625/lar.2015.31.3.101

[55] Xiang F, Peng L, Yin Z, Jia R, Hu Z, Li Z, et al. Acute and subchronic toxicity as well as evaluation of safety pharmacology of Galla chinensis solution. Journal of Ethnopharmacology. 2014;**162**:181-190. DOI: 10.1016/j.jep.2014.12.021

[56] Kim B, Choi M, Lee H, Lee S, Noh K, Kwon S, et al. Photoprotective potential of penta-O-galloyl-b-D-glucose by targeting NF-jB and MAPK signaling in UVB radiation-induced human dermal fibroblasts and mouse skin. Molecules and Cells. 2015;**38**:982-990. DOI: 10.14348/molcells.015.0169

[57] Yong Z, Bao-en W, Shu-wen Z, Shu-min Y, Hong W, Ai-min R, et al. Isolation of antifungal compound from *Paeonia suffruticosa* and its antifungal mechanism. Chinese Journal of Integrative Medicine. 2015;**21**:211-216. DOI: 10.1007/s11655-014-1805-7

[58] Jin F, Ma K, Chen M, Zou M, Wu Y, Li F, et al. Pentagalloylglucose blocks the nuclear transport and the process of nucleocapsid egress to inhibit HSV-1 infection. Japanese Journal of Infectious Diseases. 2016;**69**:135-142. DOI: 10.7883/yoken.JJID.2015.137

[59] Zhao W, Haller V, Ritsch A. The polyphenol PGG enhances expression of SR-BI and ABCA1 in J774 and THP-1 macrophages. Atherosclerosis. 2015;**242**: 611-617. DOI: 10.1016/j.atherosclerosis.2015.08.025

[60] Hanhineva K, Torronen R, Bondia-Pons I, Pekkinen J, Kolehmainen M, Mykkanan H, et al. Impact of dietary polyphenols on carbohydrate metabolism. International Journal of Molecular Sciences. 2010;**11**:1365-1402

[61] Ren Y, Himmeldirk K, Chen X. Synthesis and structure–activity relationship study of antidiabetic penta-O-galloyl-D-glucopyranose and its analogues. Journal of Medicinal Chemistry. 2006;**49**:2829-2837. DOI: 10.1021/jm060087k

[62] Faried A, Kurnia D, Faried L, Usman N, Miyazaki T, Kato H, et al. Anticancer effects of gallic acid isolated from Indonesian herbal medicine, Phaleria macrocarpa (Scheff.) Boerl, on human cancer cell lines. International Journal of Oncology. 2007;**30**:605-613. DOI: 10.3892/ijo.30.3.605

[63] Pramod D, Seung J, Youn H, Sp N, Dong Y, Hyo J, et al. Tannic acid inhibits

the Jak2/STAT3 pathway and induces G1/S arrest and mitochondrial apoptosis in YD-38 gingival cancer cells. International Journal of Oncology. 2015; **47**:1111-1120. DOI: 10.3892/ijo.2015.3098

[64] Al-Halabi R, Abou Merhi R, Chakilam S, El-Baba C, Hamade E, Pietro Di F, et al. Gallotannin is a DNA damaging compound that induces senescence independently of p53 and p21 in human colon cancer cells. Molecular Carcinogenesis. 2014;**54**: 1037-1050. DOI: 10.1002/mc.22172

[65] Al-Ayyoubi S, Gali-Muhtasib H. Differential apoptosis by gallotannin in human colon cancer cells with distinct p53 status. Molecular Carcinogenesis. 2007;**46**:176-186. DOI: 10.1002/mc.20252

[66] Sayoni N, Krishnendu M, Krishna D. Tannic acid-stabilized gold nano-particles are superior to native tannic acid in inducing ROS-dependent mitochondrial apoptosis in colorectal carcinoma cells *via* the p53/AKT axis. RSC Advances. 2019;**9**(14):8025-8038. DOI: 10.1039/C9RA00808J

[67] Hu H, Zhang J, Lee J, Kim S, Lu J. Penta-O-galloyl-beta-D-glucose induces S- and G(1)-cell cycle arrests in prostate cancer cells targeting DNA replication and cyclin D1. Carcinogenesis. 2009;**30**: 818-823. DOI: 10.1093/carcin/bgp059

[68] Hee J, Hee Y, Eun J, Hyunsuk K, Bogeun K, Kwon J, et al. Suppression of E-cadherin mediates gallotannin induced apoptosis in hep G2 hepatocellular carcinoma cells. International Journal of Biological Sciences. 2014;**10**:490-499. DOI: 10.7150/ijbs.7495

[69] Seon-Mi Y, Eun-Jeong G, Ki-Wha C, Kwang-Hoon K, Hong-Sik C, Song-Ja K. Gallotannin regulates apoptosis and COX-2 expression via Akt and p38kinase pathway in human lung cancer cell line, A549. Animal Cells and Systems. 2012;**16**:366-375. DOI: 10.1080/19768354.2012.696553

[70] Eunkyung P, Hee Y, Ji H, Doek-Beom J, Arong J, Jinhong C, et al. Inhibition of myeloid cell leukemia 1 and activation of caspases are critically involved in gallotannin-induced apoptosis in prostate Cancer cells. Phytotherapy Research. 2015;**29**(8): 1225-1236. DOI: 10.1002/ptr.5371

[71] Kwon H, Kim J, Kim B, Srivastava S, Kim S. Regulation of SIRT1/AMPK axis is critically involved in gallotannin-induced senescence and impaired autophagy leading to cell death in hepatocellular carcinoma cells. Archives of Toxicology. 2018;**92**:241-257. DOI: 10.1007/s00204-017-2021-y

[72] Prashanth N, Elham H, Pallabita C, Vivek K, Sheema K, Bilal H, et al. Tannic acid induces endoplasmic reticulum stress-mediated apoptosis in prostate Cancer. Cancers. 2018;**10**:68. DOI: 10.3390/cancers10030068

[73] Sangkil N, David M, Ping Q. Tannic acid potently inhibits tumor cell proteasome activity, increases p27 and bax expression, and induces G_1 arrest and apoptosis. Cancer Epidemiology, Biomarkers & Prevention. 2001;**10**: 1083-1088

[74] Seon-Mi Y, Song-Ja K. Gallotannin (GT) regulates apoptosis and inflammation via PI3-kinase/p38 MAP kinase pathway in MDA-MB-231 human breast cancer cells. Cancer Prevention Research. 2011;**16**:200-209

[75] Ahu S, Didem T, Ayse B, Hasan V, Irfan D, Fezan S. Induction of apoptosis in human breast adenocarcinoma MCF-7 cells by tannic acid and resveratrol. African Journal of Biotechnology. 2013; **12**:1431-1437. DOI: 10.5897/AJB11.3086

[76] Bo Y, Hwa M, Yong H, Woo H. Gallic acid inhibits the growth of HeLa cervical cancer cells via apoptosis and/or necrosis. Food and Chemical Toxicology. 2010;**48**:1334-1340. DOI: 10.1016/j.fct.2010.02.034

[77] Inoue M, Sakaguchi N, Isuzugawa K, Tani H, Ogihara Y. Role of reactive oxygen species in gallic acid-induced apoptosis. Biological & Pharmaceutical Bulletin. 2000;**23**:1153-1157. DOI: 10.1248/bpb.23.1153

[78] Chen K, Hsiao Y, Kuo D, Chou M, Chu S, Hsieh Y, et al. Tannic acid-induced apoptosis and enhanced sensitivity to arsenic trioxide in human leukemia HL-60 cells. Leukemia Research. 2009;**33**:297-307. DOI: 10.1016/j.leukres.2008.08.006

[79] Zhang J, Chen D, Han D-M, Cheng Y-H, Dai C, Xiu-Jie W, et al. Tannic acid mediated induction of apoptosis in human glioma Hs 683 cells. Oncology Letters. 2018;**15**:6845-6850. DOI: 10.3892/ol.2018.8197

[80] Chai Y, Lee H, Shaik A. Penta-O-galloyl-beta-D-glucose induces G1 arrest and DNA replicative S-phase arrest independently of cyclin-dependent kinase inhibitor 1A, cyclin-dependent kinase inhibitor 1B and P53 in human breast cancer cells and is orally active against triple negative xenograft growth. Breast Cancer Research. 2010;**12**:R67. DOI: 10.1186/bcr2634

[81] Yinhui D, Shutao Y, Cheng J, Xiaohe L, Xiao G, Chong Z, et al. Involvement of autophagy induction in penta-1,2,3,4,6-O-galloyl-β-D-glucose-induced senescence-like growth arrest in human cancer cells. Autophagy. 2014; **10**:296-310. DOI: 10.4161/ auto.27210

[82] Shutao Y, Yinhui D, Jinhua L, Junxuan L, Honglo H. Penta-1,2,3,4,6-*O*-galloyl-beta-D-glucose induces senescence-like terminal S-phase arrest in human hepatoma and breast cancer cells. Molecular Carcinogenesis. 2011; **50**:592-600. DOI: 10.1002/MC.20743

[83] Krungchanuchat S, Thongtem T, Thongtem S, Pilapong C. Iron(III)-tannic molecular nanoparticles enhance autophagy effect and T_1 MRI contrast in liver cell lines. Scientific Reports. 2018; **8**:6647. DOI: 10.1038/s41598-018-25108-1

[84] Hu H, Chai Y, Wang L, Zhang J, Lee HJ, Kim SH, et al. Pentagalloylglucose induces autophagy and caspase-independent programmed deaths in human PC-3 and mouse TRAMP-C2 prostate cancer cells. Molecular Cancer Therapeutics. 2009;**8**: 2833-2843. DOI: 10.1158/1535-7163. MCT-09-0288

[85] Ammar S, Michael H, Pirkko H, Kalevi P. Inhibition of cancer cell growth by crude extract and the phenolics of *Terminalia chebula* retz. Fruit. Journal of Ethnopharmacology. 2002;**81**:327-336. DOI: 10.1016/S0378-8741(02)00099-5

[86] Fukumura D, Gohongi T, Kadambi A, Izumi Y, Ang J, Yun C, et al. Predominant role of endothelial nitric oxide synthase in vascular endothelial growth factor-induced angiogenesis and vascular permeability. Proceedings of the National Academy of Sciences. 2001;**98**:2604-2609. DOI: 10.1073/pnas.041359198

[87] Lu Y, Jiang F, Jiang H, Wu K, Zheng X, Cai Y, et al. Gallic acid suppresses cell viability, proliferation, invasion and angiogenesis in human glioma cells. European Journal of Pharmacology. 2010;**641**:102-107. DOI: 10.1016/j.ejphar.2010.05.043

[88] Al-Halabi R, Chedid MB, Merhi R, El-Hajj H, Zahr H, Schneider-Stock R, et al. Gallotannin inhibits NFκB signaling and growth of human colon cancer xenografts. Cancer Biology & Therapy. 2011;**12**:59-68. DOI: 10.4161/cbt.12.1.15715

[89] Mattew N, Hyemee K, Alexandria B, Susanne M. Polyphenolics from mango (*Mangifera indica* L.) suppress breast cancer ductal carcinoma

in situ proliferation through activation of AMPK pathway and suppression of mTOR in athymic nude mice. The Journal of Nutritional Biochemistry. 2017;**41**:12-19. DOI: 10.1016/j.jnutbio.2016.11.005

[90] Nivedita B, Hyemee K, Kimberly K, Stephen T, Susanne M. Mango polyphenolics suppressed tumor growth in breast cancer xenografts in mice: Role of the PI3K/AKT pathway and associated microRNAs. Nutrition Research. 2015;**35**:744-751. DOI: 10.1016/j.nutres.2015.06.002

[91] Pramond D, Youn J, Dong K, Young Y. Tannic acid inhibits EGFR/STAT1/3 and enhances p38/STAT1 signalling axis in breast cancer cells. Journal of Cellular and Molecular Medicine. 2017;**21**:720-734. DOI: 10.111/jcmm.13015

Section 2

Autophagy and Cell Death

Chapter 5

Autophagy and Cell Death: Antitumor Drugs Targeting Autophagy

Hai Zhang and Zhinan Chen

Abstract

Autophagy, a degradation mechanism conserved among eukaryotes, plays an important role in cellular homeostasis by maintaining nutrients and energy balance. It is not surprising that autophagy has been associated with various pathological conditions such as neurodegeneration, aging, infection, and cancer. Its roles in cancer are complex and context-dependent. In this chapter, we will give an overview of regulation of autophagy with an emphasis in cancer and summarize the recent efforts in developing cancer therapeutics targeting autophagy.

Keywords: autophagy, autophagic cell death (ACD), autophagy-related genes (ATGs), signaling transduction pathway, antitumor drugs

1. Introduction

Autophagy derives from the Greek word "auto" and "phagein," auto means "self," and phagein means "eating," so autophagy refers to a "self-eating" physiological phenomenon.

The concept was first proposed in the 1960s, when de Duve observed that intracellular components are encased in membranes to form cystic structures and transported to a small compartment, the lysosome, which is responsible for recycling to degrade these ingredients [1]. Further studies show cell components, even organelles, are transported to the lysosome for degradation by cytoplasmic vesicles. Thus, the term autophagy was used for the first time to describe this process. But not until the early 1990s, the seminal work by Yoshinori Ohsumi on *Saccharomyces cerevisiae* resolved its molecular mechanism [2]. Now it is well-known that autophagy is the mechanism which maintains cellular homeostasis; damaged organelles or misfolded proteins are digested in lysosomes through autophagy for cellular energy recycling. However, autophagy is involved in the occurrence of cancer, neurodegenerative diseases, aging, and infection under pathological status. In this chapter, we focus on the formation and biological function of autophagy in cancer and discuss the application of antitumor drugs that target autophagy-related molecules or regulate autophagic activities.

IntechOpen

2. Molecular mechanisms and biological functions of autophagy

2.1 Molecular mechanisms of autophagy

Autophagy is a lysosomal-mediated self-degradation process of intracellular components. Under normal physiological conditions, autophagy degrades damaged organelles and nonfunctional proteins in cells to provide energy and nutrients to maintain intracellular homeostasis. However, autophagy may be induced to participate in disease processes during fasting, drug interactions, nutritional deficiencies, hypoxia, or other stress reactions. Autophagy is divided into three stages: initiation, phagophore membrane formation and extension, and maturation. When the cellular sensation is subjected to external pressure, the signal-transduction molecules are activated, which regulates autophagy-related genes (ATGs) and promotes autophagy. The molecular mechanisms of the autophagy process are summarized in **Figure 1** [3].

The energy receptor, AMP-activated protein kinase (AMPK), is activated to rapidly induce the upregulation of autophagy levels when cells are under nutrient and energy deficiency, protein accumulation, and stress. Autophagy induction is mainly achieved by the interaction between the Unc-51 like autophagy activating kinase 1 (ULK1) complex (including Atg1/ULK1, Atg17/FIP200, and Atg13) and mammalian target of rapamycin complex 1 (mTORC1) [4]. First, mTORC1 is activated when cellular energy is sufficient, which then phosphorylates ULK1 and Atg13 to inhibit autophagy. In contrast, mTORC1 activity is inhibited when cellular energy is deficient, and the resultant dephosphorylated Atg13 forms a complex with ULK1 and interacts with FIP200 to initiate autophagy [5]. The initiation of autophagy is closely related to the Class III PI3K (Vps34)-Atg6/Beclin1 complex [6, 7]. The Beclin1-PI3K complex recruits Atg12-Atg5 and Atg16L multimers and Atg8/microtubule-associated protein light chain 3 (LC3), which promotes the stretching and extension of autophagosome membranes [8]. The expansion of autophagosomes mainly depends on two ubiquitin-like conjugation systems: Atg12

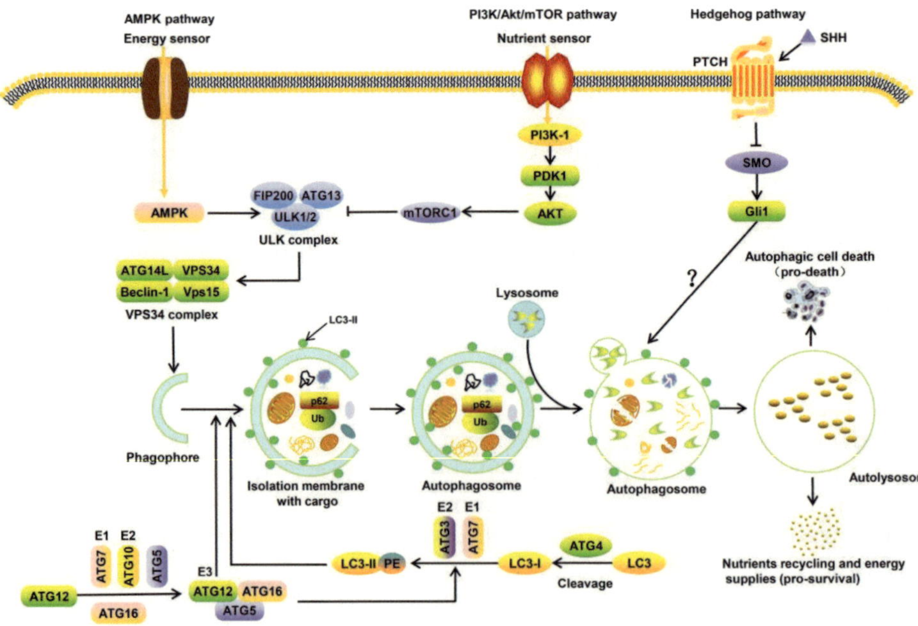

Figure 1.
Schematic diagram of the molecular mechanisms of autophagy.

binding and LC3-modification processes [9]. Atg12 binding is a ubiquitin-like process that requires the participation of ubiquitin-activating enzymes, E1 and E2. Atg12 is firstly activated by Atg7 (E1-like enzyme) and is then transported to Atg5 through Atg10 (E2-like enzyme), which subsequently binds to Atg16 to form a multibody complex and participates in the expansion of the autophagosome [10–12]. The LC3-modification process also requires the participation of ubiquitin-activating enzymes, E1 and E2. After the formation of the LC3 precursor, it is processed into cytosolic soluble LC3-I by Atg4 and is then covalently linked to phosphatidylethanolamine (PE) by Atg7 (E1-like enzyme) and Atg3 (E2-like enzyme) to form a fat-soluble LC3-II-PE that participates in the extension of the autophagosome membrane. LC3-II binds to the newly formed autophagosome membrane until the formation of the autolysosome. Therefore, LC3-II is often used as a marker for autophagy [11, 13]. After formation, autophagosomes fuse with lysosomes through the microtubule cytoskeleton under the action of the endosomal-sorting complex required for transport (ESCRT), as well as through monomeric GTPase (RabS), to form autolysosomes. The autophagy-receptor protein, P62/SQSTM1, recognizes and binds to autophagy-substrate proteins by binding to the ubiquitin-associated (UBA) domain, either dependently or independently. P62/SQSTM1 also anchors to the autophagosome membrane through the LC3/Atg8 interaction region (LIR); ultimately fusion between autophagosomes and lysosomes leads to P62/SQSTM1 entering into autolysosomes, together with the corresponding substrates and their degradation [14].

In addition to the direct involvement of the Atg gene in autophagy, some important signaling pathways also regulate autophagy. Among these pathways, the PI3k/Akt/mTOR pathway is a classical signal-transduction pathway associated with autophagy. Under nutrient-rich conditions, PI3k/Akt signaling activates the downstream mammalian target of rapamycin (mTOR) to form the mTOR complex consisting of ULK1/2, mATG13, FIP200, and Atg101. Once the mTOR complex is formed, phosphorylation of ULK1 and Atg13 results in ULK1 inactivation, which suppresses autophagy. In the absence of nutrients, the LKB/AMPK pathway is activated, and the binding of mTORC1 to ULK is blocked, thereby suppressing mTOR activity. Subsequently, ULK1 phosphorylates Atg13 and FIP200 to form autophagosomes, ultimately inducing autophagy [15]. The AMPK pathway is a cellular energy sensor, and it regulates the occurrence of autophagy. In the absence of nutrients, AMPK senses the changes in AMP levels and is activated, thereby phosphorylating tuberous sclerosis 2 (TSC2) and aggravating the inhibition of Rheb GTPase by TSC1/2, which ultimately inhibits mTOR activity and induces autophagy. In addition, AMPK activates ULK1 through phosphorylation, thereby promoting autophagy. Under nutrient-rich conditions, mTOR phosphorylates ULK1 to prevent AMPK from activating ULK1, which ultimately suppresses ULK1 activity and thereby inhibits autophagy [16]. The Hedgehog (Hh) signaling pathway also regulates autophagy. After Sonic hedgehog (SHh) in the Hh signaling pathway senses signaling stimulation, it transfers and binds to PATCHED-1 (PTCH) receptor, thereby releasing inhibition of the G-protein-coupled receptor (GPCR)-like protein Smoothened (Smo). The activation of Smo ultimately leads to the activation of the downstream Gli transcription factor. Gli has three homologues, of which Gli1 is involved in autophagy and inhibition of Gli1 induces autophagy. However, the mechanism by which Gli1 induces autophagy has not been well clarified [17, 18].

2.2 Autophagic cell death (ACD)

Programmed cell death is an orderly process, and, once this process is initiated, it proceeds automatically. Programmed cell death can be classified into at least three types: type I (apoptosis), type II (ACD), and type III (necrosis or lysosomal death). Type I programmed cell death is accompanied with typical morphological changes of the cells—such as cell shrinkage, chromatin condensation, and apoptotic-body formation—following the initiation of apoptosis (**Figure 2a**). The morphological features of cells undergoing type II programmed cell death are not as obvious, and only autophagic vacuoles can be observed under electron microscopy (**Figure 2b**). Necrosis is a form of cell death characterized by organelle swelling, cytoplasmic-membrane rupture, and leakage of cellular contents (**Figure 2c**). Therefore, ACD should be defined according to the following three features: (1) apoptosis is not triggered; (2) there is an increase in autophagic flux before or during cell death, rather than a mere elevation in autophagy marker; and (3) application of pharmaceutical inhibitor, like 3-MA, chloroquine, etc., or small interfering RNA (siRNA) against autophagy-related genes (Beclin1 or Atg5) inhibits cell death [19–21]. In short, ACD is a type of cell death caused by autophagy. Autophagy may also be involved in other types of cell death. Therefore, it is necessary to distinguish the difference between cell death that occurs with autophagy and cell death that is caused by autophagy. If inhibition of autophagy only changes cell morphologies but not the fate of cell death, then the death in question is cell death with autophagy. If inhibition of autophagy leads to alleviation or inhibition of cell death, then the death in question is cell death by autophagy [22].

The discovery of ACD has provided new ideas and strategies for anticancer therapy. It is well-known that autophagy plays a role as a "double-edged sword" in tumorigenesis (i.e., as a promoter and a suppressor of tumors). In terms of promoting tumorigenesis, autophagy is a protective mechanism that protects tumor cells from being killed. However, ACD breaks the protective mechanism of autophagy. Drug and hypoxia-mediated ACD converts the autophagy that originally benefits tumor-cell survival to become harmful to tumor cells and inhibits tumor-cell proliferation. In addition, there is a relationship between ACD and apoptosis, and autophagy has been considered to be essential for the occurrence of apoptosis. Autophagy usually occurs before apoptosis, with apoptosis being initiated after autophagy, which leads to accelerated cell death. Some researchers believe that the occurrence of autophagy inhibits apoptosis and, thus, protects tumor cells. In addition, apoptosis can coexist with autophagy to collectively promote cell death [23]. Therefore, regardless of the single role of ACD or the combined effect of ACD and apoptosis, the above studies provide an important theoretical basis for the development of antitumor drugs. In fact, many antitumor drugs, either newly developed ones or old drugs with a new target, share the same mechanism in anticancer therapy by targeting programmed cell death.

Since the discovery of autophagy, especially its role in tumorigenesis and tumor development, the use of autophagy to regulate the programmed cell death of tumor cells has become a useful tool for anticancer therapy. Drug-induced ACD is also a

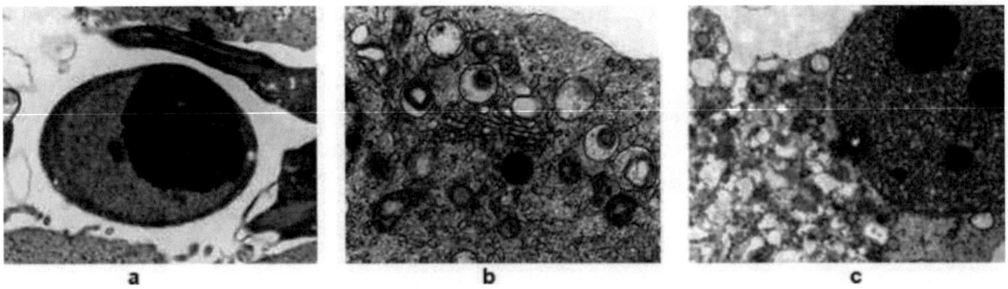

Figure 2.
Types of programmed cell death. a) apoptosis, b) autophagy and c) necrosis [22].

primary mechanism for antitumor drugs. Therefore, autophagy has been widely studied as a drug target in antitumor research. First, this new research strategy of old drugs provides ideas for autophagy-targeted drug discovery. Some repurposing drugs—such as the antimalarial drug chloroquine, the antifungal drug bafilomycin A1, and the immunosuppressant rapamycin—had not had their autophagy functions revealed when they were discovered. However, subsequent studies have revealed their other functions, which do not only induce autophagy but also inhibit tumor-cell proliferation through regulating autophagic activity. These old drugs are good candidates for being repurposed into antitumor drugs. Additionally, in the past few years with the development of omics technologies, a number of small-molecule compounds targeting autophagy molecules or autophagy pathways have been discovered by high-throughput screening methods. These drugs also inhibit tumor-cell proliferation through the ACD pathway. The following section summarizes drugs that control tumor proliferation by regulating autophagy activity in cancer stem cells (CSCs) and tumor cells, as well as their mechanisms of action.

3. Antitumor drugs that regulate autophagy to target cancer stem cells

Cancer stem cells (CSCs) are a type of cell with stem cell characteristics in tumors that exhibit strong self-renewal and proliferative ability and play an important role in tumor survival, proliferation, metastasis, and recurrence. Taken together, CSCs maintain the vitality of tumor-cell populations through self-renewal and infinite proliferation [24]. CSCs in tumor tissues in microenvironments with insufficient nutrients increase the recycling of intracellular substances by autophagy to continuously survive. Apoptosis, proliferation, and differentiation of CSCs are regulated by many factors, and autophagy and its signaling pathways play an important role in the apoptosis and differentiation of CSCs [25, 26]. A previous study has shown that the autophagy activity regulated by autophagy-related genes or autophagy pathways promotes the survival of CSCs. High Beclin1 expression promotes the survival and tumorigenicity of breast CSCs [27]. Chloroquine inhibits autophagy via the Janus kinase 2 (JAK2) and Hh pathways and kills breast and pancreatic CSCs [28, 29]. Nevertheless, another study has shown that autophagy also promotes the death of CSCs. CSCs from malignant gliomas cause cell death due to the accumulation of autophagy-related proteins and autophagosomes [30]. In addition, CSCs adapt to the changes of the microenvironment simultaneously through autophagy, and a large number of autophagosomes are found in damaged blood vessels and hypoxic parts of tumors. Autophagy induced by hypoxia-inducible factor (HIF) promotes metastasis of CD133+ pancreatic CSCs. Inhibition of autophagy reduces the viabilities of pancreatic cancer cells with stem cell characteristics and CD133+ liver CSCs under oxygen/nutrition deprivation [31]. Multi-regulation of autophagy on CSCs has prompted researchers to explore the feasibility of autophagy as a drug target for anticancer therapy. Numerous studies have demonstrated that some chemical synthetic drugs or natural extracts have a clear therapeutic effect on CSCs through regulating autophagy activity and may be used as novel antitumor treatments.

Small-molecule compounds induce programmed cell death in tumor cells through apoptosis and autophagy as their main antitumor mechanisms. After the concept of CSCs was proposed, researchers began to continuously explore small-molecule compounds and their effective targeting of CSCs. In 2009, Gupta et al. screened more than 16,000 compounds to discover that salinomycin has strong killing activity against breast CSCs and that the killing activity is >100-fold higher than that of paclitaxel [32]. Further studies have shown that salinomycin has an inhibitory effect on a variety of CSCs. Salinomycin does not only induce apoptosis in CSCs but also inhibits DNA repair of CSCs. Interestingly, the proliferation of CSCs is inhibited by the increase or decrease of autophagic activity after the action of salinomycin [33]. On the one hand, after the action of salinomycin, the increase of reactive oxygen species (ROS) activates c-Jun N-terminal kinase (JNK) and AMPK signaling pathways to induce an increase of autophagy activity; on the other hand, salinomycin also affects the process of autophagosome and lysosome fusion and reduces the level of autophagic flux. These factors are the mechanisms by which salinomycin inhibits the proliferation of CSCs [34–36]. PI3k/Akt/mTOR is a key pathway for autophagy regulation. When cells sense signs of starvation and hypoxia, they inhibit the activity of the PI3k/Akt/mTOR pathway and promote autophagy by phosphorylating Akt at amino-acid position 473 and mTOR at amino-acid position 2448. NVP-BEZ235 is a dual ATP-competitive PI3K and mTOR inhibitor, and its induced autophagy and apoptosis enhance the radiosensitivity of glioma stem cells [37]. Rottlerin, a small-molecule compound that acts as a protein kinase C-δ (PKC-δ) inhibitor, regulates the PI3k/Akt/mTOR pathway to activate autophagy. Rottlerin promotes the death of pancreatic CSCs through an endogenous apoptotic pathway after the activation of autophagy [38]. The same phenomenon is also observed in Rottlerin-treated prostate CSCs [39]. Nearly all malignant tumors have epigenetic abnormalities. Antitumor drugs targeting histone deacetylase (HDAC) are a primary focus in epigenetic research. Small-molecule inhibitors that target HDAC have a clear inhibitory effect on the proliferation of CSCs. Givinostat (ITF2357) and suberoylanilide hydroxamic acid (SAHA) are currently the most studied HDAC small-molecule inhibitors. Studies have shown that givinostat and SAHA can induce autophagy in lung and glioma CSCs, respectively, leading to cell death [40–42].

Autophagy is associated with drug resistance and metastasis of CSCs. Autophagy inhibitors or silencing of autophagy-related genes also affect the self-renewal and differentiation of CSCs or enhance the sensitivity of CSCs to chemotherapeutic drugs, thereby killing CSCs. The autophagy inhibitor, chloroquine, not only does inhibit autophagy activity of breast CSCs but also reduces the cell number and metastasis of breast CD44+/CD24−/low CSCs; this effect is damaging to the mitochondrial membrane structure, which leads to decreased cytochrome-c activity and increases oxidative stress. This process also leads to increased expression of the double-stranded DNA damage marker, γ-H2AX, which reduces the repair capacity of double-stranded DNA breaks, thereby inhibiting the proliferation of breast CSCs [43]. Drug resistance is a challenge for antitumor therapy. The small-molecule compound, baicalein competitively binds to the GTPase SAR1B protein, guanosine triphosphate, which is required for autophagy and inhibits autophagy activity, thereby selectively enhancing the sensitivity of mice liver CD133+ tumor-initiating stem cell-like cells (TICs) to mTORC1 inhibitors [44]. The autophagy inhibitor, quinacrine, which can pass through the blood-brain barrier, also increases the sensitivity of glioblastoma stem-like cells (GSCs) to the chemotherapeutic drug, temozolomide (TMZ). The combination of quinacrine and TMZ promotes iron-dependent cell death (ferroptosis) in GSCs [45]. The bromodomain and extra-terminal domain (BET) inhibitor, JQ1, is ineffective for drug-resistant CD34+ CD8- leukemia stem cells (LSCs) and hardly induces apoptosis in drug-resistant LSCs. However, JQ1 increases the expression of autophagy-related molecules, such as Beclin1 and LC3-II, through the AMPK-ULK1 pathway to promote autophagosome formation. The AMPK inhibitor, compound C, and AMPKα siRNA inhibit autophagy to further promote the apoptosis of drug-resistant CD34+CD8− LSCs [46].

In addition to small-molecule compounds, plant extracts also have an inhibitory effect on CSCs. Curcumin is a chemical component extracted from the roots

Drugs	Cancer stem cells	Autophagy activity	Mechanism of autophagy regulation	References
Chloroquine	Breast	Autophagy inhibition	Jak2, DNMT1	[27, 42]
	Pancreatic	Autophagy unchanged	CXCR4	[28]
Salinomycin	Breast	Autophagy increasing	ROS	[33]
	Breast	Autophagy inhibition	Apoptotic pathway	[35]
	Colon	Autophagy induction	ROS	[33]
NVP-BEZ235	Gliomas	Autophagy induction	PI3k/Akt/mTOR	[36]
Rettlerin	Pancreatic	Autophagy induction	Apoptotic pathway	[37]
	Prostate	Autophagy induction	Apoptotic pathway	[38]
Givinostat	Lung	Autophagy induction	HDAC	[39]
	Glioblastoma	Autophagy inhibition	HDAC	[40]
SAHA	Glioblastoma	Autophagy induction	HDAC	[41]
Baicalein	Mice liver tumor	Autophagy inhibition	GTPase SAR1B protein	[43]
Quinacrine	Glioblastoma	Autophagy inhibition	Lipid peroxides	[44]
JQ1	Leukemia	Autophagy induction	AMPK-ULK1 pathway	[45]
Curcumin	Colon	Autophagy induction	DCLK1	[46]
Resveratrol	Breast	Autophagy induction	Wnt/β-catenin pathway	[47]
	Colon adenocarcinoma	Autophagy induction	Wnt pathway	[48]
Bitter-melon fruit extracts	Colon and progenitor cells	Autophagy induction	AMPK pathway	[49]

Abbreviations: Jak2: Janus kinase 2; DNMT1: DNA (cytosine-5)-methyltransferase 1; CXCR4: C-X-C chemokine receptor type 4; ROS: reactive oxygen species; HDAC: histone deacetylases; DCLK1: Doublecortin like kinase 1; ULK1: Unc-51 like autophagy activating kinase 1

Table 1.
Antitumor drugs that target cancer stem cells by regulating autophagy.

of some plants in the Zingiberaceae and Araceae families. Among the numerous biological activities of curcumin, its antitumor activity has aroused the attention of tumor biologists. Studies have shown that curcumin has an inhibitory effect on liver and ovarian CSCs. Autophagy induced by curcumin treatment in the DCLK1+ colon CSCs inhibits cell proliferation via apoptosis [47]. Resveratrol not only does inhibit the proliferation of breast CSCs by autophagy [48] but also increases the trans-differentiation of colon CSCs into endothelial cells [49]. Bitter-melon whole fruit (BMW) and skin (BMSk) extracts significantly inhibit the proliferation and colonization of colon CSCs, and the BMW-treated colon CSCs also cause the upregulated expression of LC3-II, Beclin1, Atg7, and Atg12 to promote autophagy [50]. **Table 1** summarizes the recent progress of mechanism, targets, and references of drugs in anticancer stem cells.

4. Antitumor drugs that regulate autophagy to target tumor cells

In the processes of tumorigenesis and tumor progression, autophagy has dual roles, and inhibition or promotion of autophagy is related to tumorigenesis. Some important signaling pathways—such as PI3K/Akt/mTOR, MAPK, JNK, and Hh pathways—do not only regulate tumor-cell growth and proliferation but also affect tumorigenesis by regulating autophagic activity. In addition, autophagy-related molecules—such as Atg5, Atg7, and Beclin1—are associated with tumorigenesis. Therefore, research in recent years regarding autophagy-targeted antitumor drugs includes not only autophagy-related signaling pathways but also autophagy-related molecules. In this section, antitumor drugs, like compounds and nature extracts, which target the signaling pathway (**Table 2**) or autophagy-related molecules (**Table 3**), are introduced, and their antitumor mechanisms are discussed.

Drugs	Cancer cell	Autophagy activity	Mechanism of autophagy regulation	References
Metformin	Myeloma	Autophagy induction	AMPK	[50]
	Rat liver carcinoma	Autophagy inhibition	AMPK, p38/MAPK	[51]
Cannabinoids	Pancreatic cancer	Autophagy induction	AMPK	[52]
Safingol	Colon carcinoma	Autophagy induction	PI3K/Akt/mTOR	[54]
Compound 6	Lung carcinoma	Autophagy induction	PI3K/Akt/mTOR	[55]
Cabazitaxel	Lung carcinoma	Autophagy induction	PI3K/Akt/mTOR	[56]
Ophiopogon japonicus	Lung carcinoma	Autophagic cell death	PI3K/Akt/mTOR	[57]
Capsaicin	Nasopharyngeal cancer cell	Autophagy induction	PI3K/Akt/mTOR	[58]
NVP-BEZ235	T-cell acute lymphoblastic leukemia	Autophagy induction	PI3K/Akt/mTOR	[59]
CYT-Rx20	Breast cancer	Autophagy induction	MAPK	[60]
AJ-5	Melanoma	Autophagic cell death	MAPK	[61]
Arsenic-trioxide	Glioma	Autophagic cell death	PI3K/Akt, ERK1/2	[62]
Fangchinoline	Liver cancer	Autophagic cell death	p53/sestrin2/AMPK	[63]

Drugs	Cancer cell	Autophagy activity	Mechanism of autophagy regulation	References
PEITC	Liver cancer	Autophagy induction	p53	[64]
Bafilomycin A1	B-cell acute lymphoblastic leukemia cell	Autophagy inhibition	Autophagosome-lysosome fusion inhibition	[66]
DQ661	Melanoma	Autophagy inhibition	Autophagosome-lysosome fusion inhibition	[67]
lys05	Glioblastoma	Autophagy inhibition	Autophagosome-lysosome fusion inhibition	[68]
Itraconazole	Breast and liver cancer	Autophagic cell death	Hh	[17]
GANT61	Breast and liver cancer	Autophagic cell death	Hh	[17]
HhAntag	Embryonal rhabdomyosarcoma	Autophagy induction	Hh	[69]
Sonidegib	Mantle cell lymphoma	Autophagy inhibition	Hh	[70]
SAHA	Colon and liver cancer	Autophagy induction	HDAC inhibitor	[71]
Panobinostat	Colon cancer	Autophagy induction	HDAC inhibitor	[72, 73]
MGCD0103	B-cell chronic lymphocytic leukemia	Autophagy inhibition	HDAC inhibitor	[74]

Abbreviations: AMPK: 5′ AMP-activated protein kinase; MAPK: mitogen-activated protein kinase; PEITC: phenethyl isothiocyanate; SAHA: Suberoylanilide hydroxamic acid; HDAC: histone deacetylases; Hh: Hedgehog

Table 2.
Antitumor drugs that target cancer cells by regulating autophagy.

4.1 Autophagy pathways

4.1.1 Drugs that target AMPK pathway

AMPK is a metabolism and energy receptor of cells. The intracellular balance of energy and metabolism in tumor cells is often chaotic, leading to changes in AMPK activity to further cause activity changes of a series of downstream signaling pathways, which affect the autophagy process. In addition to its therapeutic effect on type II diabetes, metformin has also been found to have a therapeutic effect on tumors. Wang et al. has shown that metformin targets the molecules of AMPK/mTORC1 and mTORC2 pathways in the tumor cells of multiple myeloma. Activation of AMPK signaling inhibits mTORC1 and mTORC2 to further induce autophagy and inhibit myeloma cell proliferation [51]. In contrast, metformin also inhibits autophagy and promotes apoptosis through AMPK and p38/MAPK signaling pathways in the glucose-deprived H4IIE rat hepatocellular-carcinoma cell line [52]. Cannabinoids have a variety of biological activities, of which tumor inhibition has aroused attention from researchers. AMPK pathways activate autophagy after ROS-dependent activation in cannabinoid-treated pancreatic cancer cells, while ROS promotes GAPDH nuclear translation, leading to reduced glycolysis and NADH accumulation to block the Krebs cycle after blocking the respiratory-chain function. Simultaneously with the activation of the AMPK pathway, cannabinoids also inhibit the Akt/c-Myc pathway, resulting in the reduction of pyruvate kinase isoform M2 (PKM2) activity, glycolysis, and glutamine uptake. Therefore, cannabinoids inhibit the proliferation of pancreatic cancer cells through autophagy and cellular-metabolic pathways regulated by the AMPK pathway [53].

4.1.2 Drugs that target PI3K/Akt/mTOR pathway

The PI3K/Akt/mTOR pathway is directly involved in tumor-cell proliferation, and small-molecule drugs or natural extracts act on the PI3K/Akt/mTOR pathway to regulate tumor cell proliferation through autophagy. Specifically, 3-MA, LY294002, and wortmannin are classic inhibitors of the PI3K/Akt/mTOR pathway to suppress autophagy, which have killing effects on various tumor cells [54]. In addition, many compounds that regulate autophagy and inhibit tumor cells via the PI3K/Akt/mTOR pathway have been discovered in recent years. Protein kinase C and the sphingosine kinase inhibitor, Salingol, inhibit the phosphorylation level of key molecules of the PI3K/Akt/mTOR pathway—such as Akt, p70S6k, and rS6—to induce ACD in colon cancer cells [55]. A novel hybrid of the 3-benzyl coumarin seco-B-ring derivative and phenylsulfonylfuroxan (compound 6) reduces the phosphorylation levels of mTOR-S2448, Akt-S373, and the downstream p70s6K-S371 and 4EBP1-pT45 proteins in a concentration-/time-dependent manner to induce autophagy and apoptosis of A549 cells [56]. The antiprostate-cancer drug, cabazitaxel, also inhibits proliferation of A549 cells through autophagy which is regulated by the PI3K/Akt/mTOR pathway [57]. *Ophiopogon japonicus* is a traditional medicinal plant, and its main components are flavonoids and steroidal saponins. A previous study has shown that *O. japonicus* inhibits PI3K/Akt/mTOR signaling to induce ACD in A549 lung cancer cells [58]. Similarly, the inhibitory effect of capsaicin in the nasopharyngeal carcinoma cell line, NPC-TW01, is also achieved by autophagy regulated by the PI3K/Akt/mTOR pathway [59]. Specifically, mTORC1 is a key signaling molecule in the PI3K/Akt/mTOR pathway. Some small-molecule drugs, such as rapamycin and NVP-BEZ235, inhibit mTORC1 activity and induce autophagy and have a killing effect on T-cell acute lymphoblastic leukemia cells [60].

4.1.3 Drugs that target the MAPK pathway

The MAPK pathway plays an important role in the process of tumor-cell proliferation, growth, apoptosis, and cell-to-cell functional synchronization. The MAPK family members, ERK, p38 MAPK, JNK, and ERK5 are also involved in the regulation of autophagy. CYT-Rx20, a derivative of β-nitrostyrene, inhibits proliferation in the breast cancer cell lines, MDA-MB-231 and MCF-7 [61]. A new binuclear palladacycle complex, AJ-5, does not only induce apoptosis in melanoma cells but also inhibit Akt/mTOR activity and activate p38 and ERK1/2 MAPK signaling to induce ACD, with dual roles in apoptosis and ACD to inhibit the proliferation of melanoma cells [62]. Similar to AJ-5, arsenic-trioxide and ionizing-radiation combination treatment on glioma cells also inhibits Akt/mTOR activity and activates ERK1/2 MAPK signaling to promote ACD [63].

4.1.4 Drugs that target the p53 pathway

p53 is a tumor-suppressor protein that regulates the expression of multiple genes involved in apoptosis, growth inhibition, and DNA repair. Studies have shown that the regulation of autophagy by p53 is bidirectional depending on its localization. Cytoplasmic p53 inhibits autophagy in a transcriptional-independent manner, while nuclear p53 enhances autophagy by transactivation target gene.

Drugs	Cancer cell	Autophagy activity	Mechanism of autophagy regulation	Refs.
SBI-0206965	Lung cancer cell	Autophagy inhibition	ULK1 inhibitor	[75]
MRT67307	Mouse embryonic fibroblast	Autophagy inhibition	ULK1 inhibitor	[76]
MRT68921	Mouse embryonic fibroblast	Autophagy inhibition	ULK1 inhibitor	[76]
LYN-1604	Breast cancer cell	Autophagy induction	ULK1 inhibitor	[77]
SAR405	Renal tumor	Autophagy inhibition	Vps34 inhibitor	[78]
Spautin-1	Cervical cancer	Autophagy inhibition	Vps34 inhibitor	[80]
S130	Colon cancer	Autophagy inhibition	Atg4B inhibitor	[81]
NSC185058	Saos-2 osteosarcoma cell	Autophagy inhibition	Atg4B inhibitor	[82]
Obatoclax	Lymphoblastic leukemia	Autophagic cell death	Bcl-2 inhibitor	[87]
ABT-737	Cervical cancer	Autophagy induction	Bcl-2 inhibitor	[86]
ABT-737	Colon cancer	Autophagy inhibition	Bcl-2 inhibitor	[87]
gossypol	Glioma	Autophagic cell death	Bcl-2 inhibitor	[88, 89]
Verteporfin	Prostate cancer cell	Autophagy inhibition	p62/SQSTM1 inhibitor	[90]

Table 3.
Small molecular inhibitors of autophagy-related molecules.

P53 in human hepatocellular-carcinoma cells treated with fangchinoline is transported from the cytoplasm to the nucleus through nuclear translocation and selectively activates sestrin2 to initiate autophagy and promote ACD [64]. Numerous studies have shown that cruciferous-vegetable-derived phenethyl isothiocyanate (PEITC) reactivates the normal activity of mutant p53 molecules in tumor cells and has an antitumor effect. PEITC has growth-inhibitory activity on tumor cells expressing p53R175H and restores the wide-type conformation and transcriptional activity of p53. In addition, PEITC enables p53R175H tumor cells to be sensitive to proteasome and autophagy-mediated degradation processes. Analysis of the mechanism showed that PEITC activates the classical p53 downstream target gene, causing tumor cells to arrest in the S phase and G2/M phase of the cell cycle and to undergo apoptosis [65].

4.1.5 Drugs that target the autophagosome-lysosome fusion pathway

The fusion of autophagosomes and lysosomes forms autolysosomes, which degrade the encapsulated materials. The degradation activity can be evaluated by autophagic flux. A previous study has shown that antimalarial drugs—such as quinine, chloroquine, and its derivatives—block the fusion processes of autophagosomes and lysosomes to reduce the autophagic flux and kill tumor cells by inhibiting autophagy activity [66]. Bafilomycin A1 is a vesicular H^+-ATPase proton-pump inhibitor that inhibits autolysosome activity to promote tumor cell killing by inhibiting the activity of vacuolar V-ATPases [67]. In addition to the above drugs, studies in recent years have shown that some small-molecule drugs, such as lys05 and DQ661, share similar antitumor mechanisms [68, 69].

4.1.6 Drugs that target the Hh pathway

The Hh pathway plays an important role in embryonic development and tissue regeneration. Abnormal expression of SMO, PTCH, and Gli1 molecules in the Hh pathways is associated with tumorigenesis. The Gli1 small-molecule inhibitor, GANT61, does not only inhibit the proliferation of human hepatocellular-carcinoma cells but also synergize with itraconazole to kill breast cancer cells through the ACD pathway [18]. HhAntag, a small-molecule inhibitor of downstream element of Hh pathway, inhibits embryonal rhabdomyosarcoma (ERMS) proliferation through autophagy induction [70]. SMO antagonist, Sonidegib, selectively targets cell migration and adhesion of mantle cell lymphoma (MCL) and suppresses the proliferation of MCL via autophagy inhibition [71].

4.1.7 Drugs that target the epigenetic pathways

Epigenetic modification can regulate the occurrence of autophagy, including histone acetylation and DNA methylation, which play an important role in regulating the biological function of autophagy. Inhibition of HDAC activity to induce cell death has been the research strategy of antitumor drugs. Development of small-molecule inhibitors targeting HDAC has become a primary focus in this field. The small-molecule inhibitor, SAHA, does not only induce a killing effect on CSCs but also promote colon cancer and liver cancer cell death by FoxO1-dependent autophagy [72]. Panobinostat (LBH589) induces autophagy in a tumor-suppressor death-associated protein kinase (DAPK)-dependent manner and inhibits colon cancer cell proliferation. The combination of panobinostat and sorafenib significantly improves the therapeutic outcomes of liver cancer treatment [73, 74]. MGCD0103 not only induces apoptosis in B-cell chronic lymphocytic leukemia (CLL) but also inhibits tumor-cell protective autophagy and promotes CCL cell death through the activation of the PI3K/AKT/mTOR signaling molecules and caspase pathways [75].

4.2 Small molecular inhibitors of autophagy-related molecules

4.2.1 Atg1/ULK1

The autophagy-related gene, Atg1 homologue, ULK1, is an unc-51-like serine/threonine kinase that plays an important role in initiating autophagy. When cells are under nutrient deficiency, hypoxia, and other stresses, the upstream molecules of ULK1, mTORC1, and AMPK are activated, and the phosphorylation of ULK1 and Atg13 molecules regulates autophagy. In recent years, ULK1 has been a popular molecule in the research of autophagy-targeted drugs. ULK1 not only is a promoter of autophagy but is also an important kinase, which is more favorable for the study of small-molecule inhibitors. The high-throughput screening of Egan et al. [76] has shown that the small-molecule compound, SBI-0206965, inhibits ULK1 activity through the mTOR pathway, leading to a decrease in autophagy and inhibition of A549 cell proliferation. Unlike SBI-0206965, MRT67307 and MRT68921 inhibit both ULK1 and ULK2 activities and block autophagy [77]. The small-molecule compound, LYN-1604, and ULK1 reactive proteins—ATF3, RAD21, and caspase 3—activate autophagy to inhibit the proliferation of MDA-MB-231 triple-negative breast cancer cells [78].

4.2.2 Vps34

Vps34 is a Class III phosphoinositide-3 kinase (PI3K-III), and its 3-phosphophosphatidylinositol (PI3P), which catalyzes the formation of the substrate phosphatidylinositol (PI), is necessary for autophagosome formation. The small-molecule inhibitor, SAR405, inhibits the catalytic activity of PIK3C3/Vps34, blocks the transport of vesicles from late endosomes to lysosomes, and inhibits autophagy, which suppresses tumor-cell proliferation [79]. In addition, PIK-III and Vps34-IN1 are also Vps34 inhibitors, which inhibit autophagy and have killing effects on tumor cells [80]. Spautin-1 is a potent and specific small-molecule inhibitor. It inhibits two ubiquitin-specific peptidases, USP10 and USP13, and promotes ubiquitination and degradation of the Vps34-PI3K complex, thereby inhibiting autophagy [81].

4.2.3 Atg4B

Atg4 is a key molecule in autophagy that cleaves the C-terminal arginine of Atg8/LC3 to expose the C-terminal glycine residue for covalent attachment to phosphatidylethanolamine (PE). This process induces Atg8/LC3-PE to be anchored to the membrane of autophagosomes to regulate autophagy. Fu et al. [82] identified a novel ATG4B small-molecule inhibitor, S130, through in silico screening and in vitro high-throughput screening system for fluorescence resonance-energy transfer and showed that S130 did not affect the autophagosome function and autophagosome fusion with lysosomes. Instead, S130 inhibited the delipidation process of LC3-PE and ultimately blocked autophagy. An antitumor study has shown that S130 effectively inhibits the growth of colon cancer cells, and nutrient deficiency further enhances the antitumor effect of S130; however, Atg4 overexpression partially counteracts the tumor-cell death process caused by S130. Similar to S130, NSC185058 is also a newly discovered small-molecule inhibitor that acts on the Atg4B target and inhibits autophagy after LC3B lipidation, leading to the death of the Saos-2 osteosarcoma cells [83].

4.2.4 B-cell lymphoma/leukemia-2 (Bcl-2)

Bcl-2 is an antiapoptotic protein containing multiple Bcl-2 homology (BH) domains. Under starvation conditions, activation of JNK1 phosphorylates Bcl-2 protein and causes the separation between Bcl-2 and Beclin1, thereby inducing autophagy [84]. In addition, multiple Bcl-2 homology 3 (BH3) mimetics can disrupt the interaction of Bcl-2/xl with Beclin1 to induce autophagy by competing for the BH3 domains [7]. Obatoclax (GX15-070) is a pan-Bcl2 inhibitor that induces apoptosis in various tumor cells. It also induces ACD in acute lymphoblastic leukemia cells in an Atg-dependent manner [85]. ABT-737, as a BH analogue, binds to the BH3 binding groove of Bcl-2 and Bcl-xl. A study by Maiuri et al. [86] has shown that the degree of polymerization of Beclin1, Bcl-2, and Bcl-xl is significantly reduced in ABT-737-treated cells, which enhances the autophagy level. However, ABT-737 also enhances the sensitivity of colon cancer cells to the chemotherapeutic drug, ixazomib, through downregulation of Mcl-1 and autophagy inhibition [87]. Similarly, gossypol is a small-molecule inhibitor of Bcl-2. Under natural conditions, gossypol is a mixture of (+)-gossypol and (−)-gossypol, and compared with (+)-gossypol, (−)-gossypol has higher antitumor activity. Gossypol and (−)-gossypol induce autophagy by inhibiting the response of Beclin1 and Bcl-2 in different tumor cells.

They can inactivate cytoprotective autophagy to cause tumor cells to escape [88], promote pro-survival autophagy, and initiate ACD [89].

4.2.5 p62

p62, also known as SQSTM1 protein, plays an autophagy and apoptotic role in tumor cells. The LIR domain of four domains of the p62 protein is responsible for binding to the autophagy-receptor protein, Atg8/LC3, while the UBA domain can recruit ubiquitinated proteins to mediate autophagic degradation. Verteporfin is an inhibitor of p62 and has no effect on cell growth under normal conditions. However, verteporfin reduces the survival of MCF-7 cells during nutrient deprivation. In vivo experiments have shown that verteporfin also inhibits tumor formation frequency of PC-3 prostate cancer cells in a xenograft model [90].

In addition to small-molecule drugs and natural extracts, some macromolecular-antibody drugs and noncoding miRNAs (miR)—such as anti-EGFR monoclonal antibody panitumumab, anti-20 monoclonal antibody rituximab, miR-22, and miR-101—have been reported to inhibit tumors in recent years through regulating autophagy activity [91, 92]. Although the emergence of autophagy provides a new idea for the development of antitumor drugs, anticancer treatments still encounter many challenges. Screening for drugs that target autophagy activity in cells may be the solution to some of these problems in the future.

5. Conclusion

The molecular mechanism and biological function of autophagy are now basically clear, and Janus role of autophagy determines that it plays an important role in the antitumor process. Some compounds, plant extracts killing tumor cells through the regulation of autophagy activity, especially induced autophagic cell death has also become an important strategy against tumor. However, there are still many obstacles to overcome in order to develop autophagic drugs; for example, there is a lack of specific biomarkers to distinguish autophagic cell death from other types of cell death. There is also a need to clarify which type of autophagy, cytoprotective, or cytotoxic should be targeted. Moreover it is difficult but important to determine to what extent autophagy should be induced before the cells reach the point of no return and undergo autophagic cell death.

Author details

Hai Zhang and Zhinan Chen*
Department of Cell Biology, National Translational Science Center for Molecular Medicine, The Fourth Military Medical University, Xi'an, China

*Address all correspondence to: znchen@fmmu.edu.cn

References

[1] de Duve C. The lysosome turns fifty. Nature Cell Biology. 2005;**7**(9):847-849. DOI: 10.1038/ncb0905-847

[2] Noda T, Suzuki K, Ohsumi Y. Yeast autophagosomes: De novo formation of a membrane structure. Trends in Cell Biology. 2002;**12**(5):231-235. DOI: 10.1016/S0962-8924(02)02278-X

[3] Cicchini M, Karantza V, Xia B. Molecular pathways: Autophagy in cancer—a matter of timing and context. Clinical Cancer Research. 2015;**21**(3):498-504. DOI: 10.1158/1078-0432.CCR-13-2438

[4] Hosokawa N, Hara T, Kaizuka T, et al. Nutrient-dependent mTORC1 association with the ULK1-Atg13-FIP200 complex required for autophagy. Molecular Biology of the Cell. 2009;**20**:1981-1991. DOI: 10.1091/mbc.e08-12-1248

[5] Kamada Y, Funakoshi T, Shintani T, et al. Tor-mediated induction of autophagy via an Apg1 protein kinase complex. The Journal of Cell Biology. 2000;**150**:1507-1513. DOI: 10.1083/jcb.150.6.1507

[6] Liang C, Lee JS, Inn KS, et al. Beclin1-binding UVRAG targets the class C Vps complex to coordinate autophagosome maturation and endocytic trafficking. Nature Cell Biology. 2008;**10**:776-787. DOI: 10.1038/ncb1740

[7] Kang R, Zeh HJ, Lotze MT, et al. The Beclin 1 network regulates autophagy and apoptosis. Cell Death and Differentiation. 2011;**18**(4):571-580. DOI: 10.1038/cdd.2010.191

[8] Mehrpour M, Esclatine A, Beau I, et al. Autophagy in health and disease. 1. Regulation and significance of autophagy: An overview. American Journal of Physiology. Cell Physiology. 2010;**298**:C776-C785. DOI: 10.1152/ajpcell.00507.2009

[9] Clague MJ, Urbe S. Ubiquitin: Same molecule, different degradation pathways. Cell. 2010;**143**:682-685. DOI: 10.1016/j.cell.2010.11.012

[10] Tanida I. Autophagosome formation and molecular mechanism of autophagy. Antioxidants & Redox Signaling. 2011;**14**:2201-2214. DOI: 10.1089/ars.2010.3482

[11] Kabeya Y, Mizushima N, Ueno T, et al. LC3, a mammalian homologue of yeast Apg8p, is localized in autophagosome membranes after processing. The EMBO Journal. 2000;**19**:5720-5728. DOI: 10.1093/emboj/19.21.5720

[12] Suzuki NN, Yoshimoto K, Fujioka Y, et al. The crystal structure of plant ATG12 and its biological implication in autophagy. Autophagy. 2005;**1**:119-126. DOI: 10.4161/auto.1.2.1859

[13] Kimura S, Fujita N, Noda T, et al. Monitoring autophagy in mammalian cultured cells through the dynamics of LC3. Methods in Enzymology. 2009;**452**:1-12. DOI: 10.1016/S0076-6879(08)03601-X

[14] Pankiv S, Clausen TH, Lamark T, et al. p62/SQSTM1 binds directly to Atg8/LC3 to facilitate degradation of ubiquitinated protein aggregates by autophagy. The Journal of Biological Chemistry. 2007;**282**:24131-24145. DOI: 10.1074/jbc.M702824200

[15] Aoki M, Fujishita T. Oncogenic roles of the PI3K/AKT/mTOR axis. Current Topics in Microbiology and Immunology. 2017;**407**:153-189. DOI: 10.1007/82_2017_6

[16] Alers S, Loffler AS, Wesselborg S, et al. Role of AMPK-mTOR-Ulk1/2 in

the regulation of autophagy: Cross talk, shortcuts, and feedbacks. Molecular and Cellular Biology. 2012;**32**:2-11. DOI: 10.1128/MCB.06159-11

[17] Zeng X, Ju D. Hedgehog signaling pathway and autophagy in cancer. International Journal of Molecular Sciences. 2018;**19**:2279-2301. DOI: 10.3390/ijms19082279

[18] Wang X, Wei S, Zhao Y, et al. Anti-proliferation of breast cancer cells with itraconazole: Hedgehog pathway inhibition induces apoptosis and autophagic cell death. Cancer Letters. 2017;**385**:128-136. DOI: 10.1016/j.canlet.2016.10.034

[19] Lenardo MJ, Mcphee CK, Yu L. Autophagic cell death. Methods in Enzymology. 2009;**453**:17-31. DOI: 10.1016/S0076-6879(08)04002-0

[20] Shen HM, Codogno P. Autophagic cell death: Loch Ness monster or endangered species? Autophagy. 2011;**7**:457-465. DOI: 10.4161/auto.7.5.14226

[21] Shimizu S, Yoshida T, Tsujioka M, et al. Autophagic cell death and cancer. International Journal of Molecular Sciences. 2014;**15**:3145-3153. DOI: 10.3390/ijms15023145

[22] Kroemer G, Levine B. Autophagic cell death: The story of a misnomer. Nature Reviews. Molecular Cell Biology. 2008;**9**:1004-1010. DOI: 10.1038/nrm2529

[23] Mukhopadhyay S, Panda PK, Sinha N, et al. Autophagy and apoptosis: Where do they meet? Apoptosis. 2014;**19**:555-566. DOI: 10.1007/s10495-014-0967-2

[24] Reya T, Morrison SJ, Clarke MF, et al. Stem cells, cancer, and cancer stem cells. Nature. 2001;**414**:105-111. DOI: 10.1038/35102167

[25] Wei MF, Chen MW, Chen KC, et al. Autophagy promotes resistance to photodynamic therapy-induced apoptosis selectively in colorectal cancer stem-like cells. Autophagy. 2014;**10**:1179-1192. DOI: 10.4161/auto.28679

[26] Guan JL, Simon AK, Prescott M, et al. Autophagy in stem cells. Autophagy. 2013;**9**:830-849. DOI: 10.4161/auto.24132

[27] Gong C, Bauvy C, Tonelli G, et al. Beclin 1 and autophagy are required for the tumorigenicity of breast cancer stem-like/progenitor cells. Oncogene. 2013;**32**:2261-2272, 2271e-2272e. DOI: 10.1038/onc.2012.252

[28] Choi DS, Blanco E, Kim YS, et al. Chloroquine eliminates cancer stem cells through deregulation of Jak2 and DNMT1. Stem Cells. 2014;**32**:2309-2323. DOI: 10.1002/stem.1746

[29] Balic A, Sorensen MD, Trabulo SM, et al. Chloroquine targets pancreatic cancer stem cells via inhibition of CXCR4 and hedgehog signaling. Molecular Cancer Therapeutics. 2014;**13**:1758-1771. DOI: 10.1158/1535-7163.MCT-13-0948

[30] Jiang H, Gomez-Manzano C, Aoki H, et al. Examination of the therapeutic potential of Delta-24-RGD in brain tumor stem cells: role of autophagic cell death. Journal of the National Cancer Institute. 2007;**99**:1410-1414. DOI: 10.1093/jnci/djm102

[31] Zhu H, Wang D, Zhang L, et al. Upregulation of autophagy by hypoxia-inducible factor-1alpha promotes EMT and metastatic ability of CD133+ pancreatic cancer stem-like cells during intermittent hypoxia. Oncology Reports. 2014;**32**:935-942. DOI: 10.3892/or.2014.3298

[32] Gupta PB, Onder TT, Jiang G, et al. Identification of selective inhibitors of cancer stem cells by high-throughput screening. Cell. 2009;**138**:645-659. DOI: 10.1016/j.cell.2009.06.034

[33] Jiang J, Li H, Qaed E, et al. Salinomycin, as an autophagy modulator—A new avenue to anticancer: A review. Journal of Experimental & Clinical Cancer Research. 2018;**37**:26-39. DOI: 10.1186/s13046-018-0680-z

[34] Verdoodt B, Vogt M, Schmitz I, et al. Salinomycin induces autophagy in colon and breast cancer cells with concomitant generation of reactive oxygen species. PLoS One. 2012;7:e44132. DOI: 10.1371/journal.pone.0044132

[35] Zhu LQ, Zhen YF, Zhang Y, et al. Salinomycin activates AMP-activated protein kinase-dependent autophagy in cultured osteoblastoma cells: A negative regulator against cell apoptosis. PLoS One. 2013;**8**:e84175. DOI: 10.1371/journal.pone.0084175

[36] Yue W, Hamai A, Tonelli G, et al. Inhibition of the autophagic flux by salinomycin in breast cancer stem-like/progenitor cells interferes with their maintenance. Autophagy. 2013;**9**:714-729. DOI: 10.4161/auto.23997

[37] Wang WJ, Long LM, Yang N, et al. NVP-BEZ235, a novel dual PI3K/mTOR inhibitor, enhances the radiosensitivity of human glioma stem cells *in vitro*. Acta Pharmacologica Sinica. 2013;**34**:681-690. DOI: 10.1038/aps.2013.22

[38] Singh BN, Kumar D, Shankar S, et al. Rottlerin induces autophagy which leads to apoptotic cell death through inhibition of PI3K/Akt/mTOR pathway in human pancreatic cancer stem cells. Biochemical Pharmacology. 2012;**84**:1154-1163. DOI: 10.1016/j.bcp.2012.08.007

[39] Kumar D, Shankar S, Srivastava RK. Rottlerin induces autophagy and apoptosis in prostate cancer stem cells via PI3K/Akt/mTOR signaling pathway. Cancer Letters. 2014;**343**:179-189. DOI: 10.1016/j.canlet.2013.10.003

[40] Del BD, Desideri M, De Luca T, et al. Histone deacetylase inhibition synergistically enhances pemetrexed cytotoxicity through induction of apoptosis and autophagy in non-small cell lung cancer. Molecular Cancer. 2014;**13**:230-246. DOI: 10.1186/1476-4598-13-230

[41] Angeletti F, Fossati G, Pattarozzi A, et al. Inhibition of the autophagy pathway synergistically potentiates the cytotoxic activity of givinostat (ITF2357) on human glioblastoma cancer stem cells. Frontiers in Molecular Neuroscience. 2016;**9**:107-126. DOI: 10.3389/fnmol.2016.00107

[42] Chiao MT, Cheng WY, Yang YC, et al. Suberoylanilide hydroxamic acid (SAHA) causes tumor growth slowdown and triggers autophagy in glioblastoma stem cells. Autophagy. 2013;**9**:1509-1526. DOI: 10.4161/auto.25664

[43] Liang DH, Choi DS, Ensor JE, et al. The autophagy inhibitor chloroquine targets cancer stem cells in triple negative breast cancer by inducing mitochondrial damage and impairing DNA break repair. Cancer Letters. 2016;**376**:249-258. DOI: 10.1016/j.canlet.2016.04.002

[44] Wu R, Murali R, Kabe Y, et al. Baicalein targets GTPase-mediated autophagy to eliminate liver tumor-initiating stem cell-like cells resistant to mTORC1 inhibition. Hepatology. 2018;**68**:1726-1740. DOI: 10.1002/hep.30071

[45] Buccarelli M, Marconi M, Pacioni S, et al. Inhibition of autophagy increases susceptibility of glioblastoma stem cells to temozolomide by igniting ferroptosis. Cell Death & Disease. 2018;**9**:841-857. DOI: 10.1038/s41419-018-0864-7

[46] Jang JE, Eom JI, Jeung HK, et al. AMPK-ULK1-mediated autophagy confers resistance to BET inhibitor JQ1 in acute myeloid leukemia stem

cells. Clinical Cancer Research. 2017;**23**:2781-2794. DOI: 10.1158/1078-0432.CCR-16-1903

[47] Kantara C, O'Connell M, Sarkar S, et al. Curcumin promotes autophagic survival of a subset of colon cancer stem cells, which are ablated by DCLK1-siRNA. Cancer Research. 2014;**74**:2487-2498. DOI: 10.1158/0008-5472.CAN-13-3536

[48] Fu Y, Chang H, Peng X, et al. Resveratrol inhibits breast cancer stem-like cells and induces autophagy via suppressing Wnt/beta-catenin signaling pathway. PLoS One. 2014;**9**:e102535. DOI: 10.1371/journal.pone.0102535

[49] Pouyafar A, Rezabakhsh A, Rahbarghazi R, et al. Treatment of cancer stem cells from human colon adenocarcinoma cell line HT-29 with resveratrol and sulindac induced mesenchymal-endothelial transition rate. Cell and Tissue Research. 2019;**1**:1-12. DOI: 10.1007/s00441-019-02998-9

[50] Kwatra D, Subramaniam D, Ramamoorthy P, et al. Methanolic extracts of bitter melon inhibit colon cancer stem cells by affecting energy homeostasis and autophagy. Evidence-based Complementary and Alternative Medicine. 2013;**2013**:702869. DOI: 10.1155/2013/702869

[51] Wang Y, Xu W, Yan Z, et al. Metformin induces autophagy and G0/G1 phase cell cycle arrest in myeloma by targeting the AMPK/mTORC1 and mTORC2 pathways. Journal of Experimental & Clinical Cancer Research. 2018;**37**:63-75. DOI: 10.1186/s13046-018-0731-5

[52] Park DB. Metformin promotes apoptosis but suppresses autophagy in glucose-deprived H4IIE hepatocellular carcinoma cells. Diabetes and Metabolism Journal. 2015;**39**:518-527. DOI: 10.4093/dmj.2015.39.6.518

[53] Dando I, Donadelli M, Costanzo C, et al. Cannabinoids inhibit energetic metabolism and induce AMPK-dependent autophagy in pancreatic cancer cells. Cell Death & Disease. 2013;**4**:e664. DOI: 10.1038/cddis.2013.151

[54] Kocaturk NM, Akkoc Y, Kig C, et al. Autophagy as a molecular target for cancer treatment. European Journal of Pharmaceutical Sciences. 2019;**134**:116-137. DOI: 10.1016/j.ejps.2019.04.011

[55] Coward J, Ambrosini G, Musi E, et al. Safingol (L-threo-sphinganine) induces autophagy in solid tumor cells through inhibition of PKC and the PI3-kinase pathway. Autophagy. 2009;**5**:184-193. DOI: 10.4161/auto.5.2.7361

[56] Dong M, Ye T, Bi Y, et al. A novel hybrid of 3-benzyl coumarin seco-B-ring derivative and phenylsulfonylfuroxan induces apoptosis and autophagy in non-small-cell lung cancer. Phytomedicine. 2019;**52**:79-88. DOI: 10.1016/j.phymed.2018.09.216

[57] Huo R, Wang L, Liu P, et al. Cabazitaxel-induced autophagy via the PI3K/Akt/mTOR pathway contributes to A549 cell death. Molecular Medicine Reports. 2016;**14**:3013-3020. DOI: 10.3892/mmr.2016.5648

[58] Chen J, Yuan J, Zhou L, et al. Regulation of different components from *Ophiopogon japonicus* on autophagy in human lung adenocarcinoma A549Cells through PI3K/Akt/mTOR signaling pathway. Biomedicine & Pharmacotherapy. 2017;**87**:118-126. DOI: 10.1016/j.biopha.2016.12.093

[59] Lin YT, Wang HC, Hsu YC, et al. Capsaicin induces autophagy and apoptosis in human nasopharyngeal carcinoma cells by downregulating the PI3K/AKT/mTOR pathway. International Journal of Molecular

Sciences. 2017;**18**:1343-1358. DOI: 10.3390/ijms18071343

[60] Chiarini F, Grimaldi C, Ricci F, et al. Activity of the novel dual phosphatidylinositol 3-kinase/mammalian target of rapamycin inhibitor NVP-BEZ235 against T-cell acute lymphoblastic leukemia. Cancer Research. 2010;**70**:8097-8107. DOI: 10.1158/0008-5472.CAN-10-1814

[61] Hung AC, Tsai CH, Hou MF, et al. The synthetic beta-nitrostyrene derivative CYT-Rx20 induces breast cancer cell death and autophagy via ROS-mediated MEK/ERK pathway. Cancer Letters. 2016;**371**:251-261. DOI: 10.1016/j.canlet.2015.11.035

[62] Aliwaini S, Swarts AJ, Blanckenberg A, et al. A novel binuclear palladacycle complex inhibits melanoma growth *in vitro* and *in vivo* through apoptosis and autophagy. Biochemical Pharmacology. 2013;**86**:1650-1663. DOI: 10.1016/j.bcp.2013.09.020

[63] Chiu HW, Ho SY, Guo HR, et al. Combination treatment with arsenic trioxide and irradiation enhances autophagic effects in U118-MG cells through increased mitotic arrest and regulation of PI3K/Akt and ERK1/2 signaling pathways. Autophagy. 2009;**5**:472-483. DOI: 10.4161/auto.5.4.7759

[64] Wang N, Pan W, Zhu M, et al. Fangchinoline induces autophagic cell death via p53/sestrin2/AMPK signalling in human hepatocellular carcinoma cells. British Journal of Pharmacology. 2011;**164**:731-742. DOI: 10.1111/j.1476-5381.2011.01349.x

[65] Aggarwal M, Saxena R, Sinclair E, et al. Reactivation of mutant p53 by a dietary-related compound phenethyl isothiocyanate inhibits tumor growth. Cell Death and Differentiation. 2016;**23**:1615-1627. DOI: 10.1038/cdd.2016.48

[66] Levy J, Towers CG, Thorburn A. Targeting autophagy in cancer. Nature Reviews. Cancer. 2017;**17**:528-542. DOI: 10.1038/nrc.2017.53

[67] Yuan N, Song L, Zhang S, et al. Bafilomycin A1 targets both autophagy and apoptosis pathways in pediatric B-cell acute lymphoblastic leukemia. Haematologica. 2015;**100**:345-356. DOI: 10.3324/haematol.2014.113324

[68] Rebecca VW, Nicastri MC, Mclaughlin N, et al. A unified approach to targeting the lysosome's degradative and growth signaling roles. Cancer Discovery. 2017;7:1266-1283. DOI: 10.1158/2159-8290.CD-17-0741

[69] Mcafee Q, Zhang Z, Samanta A, et al. Autophagy inhibitor Lys05 has single-agent antitumor activity and reproduces the phenotype of a genetic autophagy deficiency. Proceedings of the National Academy of Sciences of the United States of America. 2012;**109**:8253-8258. DOI: 10.1073/pnas.1118193109

[70] Geyer N, Ridzewski R, Bauer J, et al. Different response of Ptch mutant and Ptch wildtype rhabdomyosarcoma toward SMO and PI3K inhibitors. Frontiers in Oncology. 2018;**8**:396-407. DOI: 10.3389/fonc.2018.00396

[71] Zhang H, Chen Z, Neelapu SS, et al. Hedgehog inhibitors selectively target cell migration and adhesion of mantle cell lymphoma in bone marrow microenvironment. Oncotarget. 2016;7:14350-14365. DOI: 10.18632/oncotarget.7320

[72] Zhang J, Ng S, Wang J, et al. Histone deacetylase inhibitors induce autophagy through FOXO1-dependent pathways. Autophagy. 2015;**11**:629-642. DOI: 10.1080/15548627.2015.1023981

[73] Gandesiri M, Chakilam S, Ivanovska J, et al. DAPK plays an important role in panobinostat-induced autophagy

and commits cells to apoptosis under autophagy deficient conditions. Apoptosis. 2012;**17**:1300-1315. DOI: 10.1007/s10495-012-0757-7

[74] Lachenmayer A, Toffanin S, Cabellos L, et al. Combination therapy for hepatocellular carcinoma: Additive preclinical efficacy of the HDAC inhibitor panobinostat with sorafenib. Journal of Hepatology. 2012;**56**:1343-1350. DOI: 10.1016/j.jhep.2012.01.009

[75] El-Khoury V, Pierson S, Szwarcbart E, et al. Disruption of autophagy by the histone deacetylase inhibitor MGCD0103 and its therapeutic implication in B-cell chronic lymphocytic leukemia. Leukemia. 2014;**28**:1636-1646. DOI: 10.1038/leu.2014.19

[76] Egan DF, Chun MG, Vamos M, et al. Small molecule inhibition of the autophagy kinase ULK1 and identification of ULK1 substrates. Molecular Cell. 2015;**59**:285-297. DOI: 10.1016/j.molcel.2015.05.031

[77] Petherick KJ, Conway OJ, Mpamhanga C, et al. Pharmacological inhibition of ULK1 kinase blocks mammalian target of rapamycin (mTOR)-dependent autophagy. The Journal of Biological Chemistry. 2015;**290**:11376-11383. DOI: 10.1074/jbc.C114.627778

[78] Zhang L, Fu L, Zhang S, et al. Discovery of a small molecule targeting ULK1-modulated cell death of triple negative breast cancer *in vitro* and *in vivo*. Chemical Science. 2017;**8**:2687-2701. DOI: 10.1039/c6sc05368h

[79] Pasquier B. SAR405, a PIK3C3/Vps34 inhibitor that prevents autophagy and synergizes with MTOR inhibition in tumor cells. Autophagy. 2015;**11**:725-726. DOI: 10.1080/15548627.2015.1033601

[80] Dowdle WE, Nyfeler B, Nagel J, et al. Selective VPS34 inhibitor blocks autophagy and uncovers a role for NCOA4 in ferritin degradation and iron homeostasis *in vivo*. Nature Cell Biology. 2014;**16**:1069-1079. DOI: 10.1038/ncb3053

[81] Liu J, Xia H, Kim M, et al. Beclin1 controls the levels of p53 by regulating the deubiquitination activity of USP10 and USP13. Cell. 2011;**147**:223-234. DOI: 10.1016/j.cell.2011.08.037

[82] Fu Y, Hong L, Xu J, et al. Discovery of a small molecule targeting autophagy via ATG4B inhibition and cell death of colorectal cancer cells *in vitro* and *in vivo*. Autophagy. 2018:1-17. DOI: 10.1080/15548627.2018.1517073

[83] Akin D, Wang SK, Habibzadegah-Tari P, et al. A novel ATG4B antagonist inhibits autophagy and has a negative impact on osteosarcoma tumors. Autophagy. 2014;**10**:2021-2035. DOI: 10.4161/auto.32229

[84] Wei Y, Pattingre S, Sinha S, et al. JNK1-mediated phosphorylation of Bcl-2 regulates starvation-induced autophagy. Molecular Cell. 2008;**30**:678-688. DOI: 10.1016/j.molcel.2008.06.001

[85] Heidari N, Hicks MA, Harada H. GX15-070 (obatoclax) overcomes glucocorticoid resistance in acute lymphoblastic leukemia through induction of apoptosis and autophagy. Cell Death & Disease. 2010;**1**:e76. DOI: 10.1038/cddis.2010.53

[86] Maiuri MC, Le Toumelin G, Criollo A, et al. Functional and physical interaction between Bcl-X(L) and a BH3-like domain in Beclin-1. The EMBO Journal. 2007;**26**:2527-2539. DOI: 10.1038/sj.emboj.7601689

[87] Yang L, Wan J, Xiao S, et al. BH3 mimetic ABT-737 sensitizes colorectal cancer cells to ixazomib through MCL-1 downregulation and autophagy inhibition. American Journal of Cancer Research. 2016;**6**:1345-1357

[88] Gao P, Bauvy C, Souquere S, et al. The Bcl-2 homology domain 3 mimetic gossypol induces both Beclin 1-dependent and Beclin 1-independent cytoprotective autophagy in cancer cells. The Journal of Biological Chemistry. 2010;**285**:25570-25581. DOI: 10.1074/jbc.M110.118125

[89] Voss V, Senft C, Lang V, et al. The pan-Bcl-2 inhibitor (−)-gossypol triggers autophagic cell death in malignant glioma. Molecular Cancer Research. 2010;**8**:1002-1016. DOI: 10.1158/1541-7786.MCR-09-0562

[90] Wang L, Kim D, Wise J, et al. p62 as a therapeutic target for inhibition of autophagy in prostate cancer. The Prostate. 2018;**78**:390-400. DOI: 10.1002/pros.23483

[91] Alinari L, Yu B, Christian BA, et al. Combination anti-CD74 (milatuzumab) and anti-CD20 (rituximab) monoclonal antibody therapy has *in vitro* and *in vivo* activity in mantle cell lymphoma. Blood. 2011;**117**:4530-4541. DOI: 10.1182/blood-2010-08-303354

[92] Zhang H, Tang J, Li C, et al. MiR-22 regulates 5-FU sensitivity by inhibiting autophagy and promoting apoptosis in colorectal cancer cells. Cancer Letters. 2015;**356**:781-790. DOI: 10.1016/j.canlet.2014.10.029

Chapter 6

Autophagy and Cell Death in Alzheimer's, Parkinson's and Prion Diseases

Samo Ribarič and Irina Milisav Ribarič

Abstract

Neurodegenerative brain disorders (NBD) impair brain cells' proteostasis with the accumulation of normal, mutant, misfolded or unfolded proteins in the endoplasmic reticulum (ER). The increased ER burden of these proteins elicits the unfolded protein response (UPR) and stimulates autophagy (AUT). In the short term, UPR and AUT attenuate ER's burden. With prolonged ER stress, the UPR changes from supporting cell survival to promoting apoptosis. The failure of the UPR, to meet the increased protein burden, leads to an increase in cytosolic protein accumulation that initially further stimulates AUT. Over time, the accumulated proteins in the cytosol undergo post-translational changes into toxic monomers and oligomers that repress AUT at multiple levels and promote cell death. This review describes the interlinked signalling pathways of AUT, apoptosis and necroptosis and their modulation by Alzheimer's, Parkinson's and prion diseases and outlines the pharmacological strategies for targeting AUT, apoptosis and necroptosis signalling pathways.

Keywords: Alzheimer's disease, apoptosis, autophagy, necroptosis, neurodegenerative brain disorders, Parkinson's disease, prion diseases, proteostasis

1. Introduction

1.1 Proteostasis in neurodegenerative brain disorders (NBD)

Proteostasis integrates synthesis, folding, trafficking and degradation of proteins. It is perturbed in the early stages of neurodegenerative brain disorders (NBD), before clinical manifestations [1–3]. Mutant, misfolded or unfolded proteins (P) or increased P production increases the endoplasmic reticulum (ER) protein burden in NBD such as Alzheimer's (AD), Parkinson's (PD) and prion diseases (PrD). This increased ER burden stimulates the unfolded protein response (UPR) and autophagy (AUT). The UPR response to ER stress is dichotomous [4–7]. During acute ER stress, UPR supports cell survival, by reducing ER's protein folding load and increasing ER's protein folding capacity. With prolonged ER stress, the UPR preferentially represses cell survival and triggers apoptosis. The failure of ER's stress responses (i.e. increased protein folding capacity and enhanced removal of mutant, misfolded or unfolded proteins by the UPR pathway) to attenuate the P burden leads to an increase in cytosolic P accumulation that further stimulates AUT. Over time, these P undergo post-translational changes and produce toxic monomers and

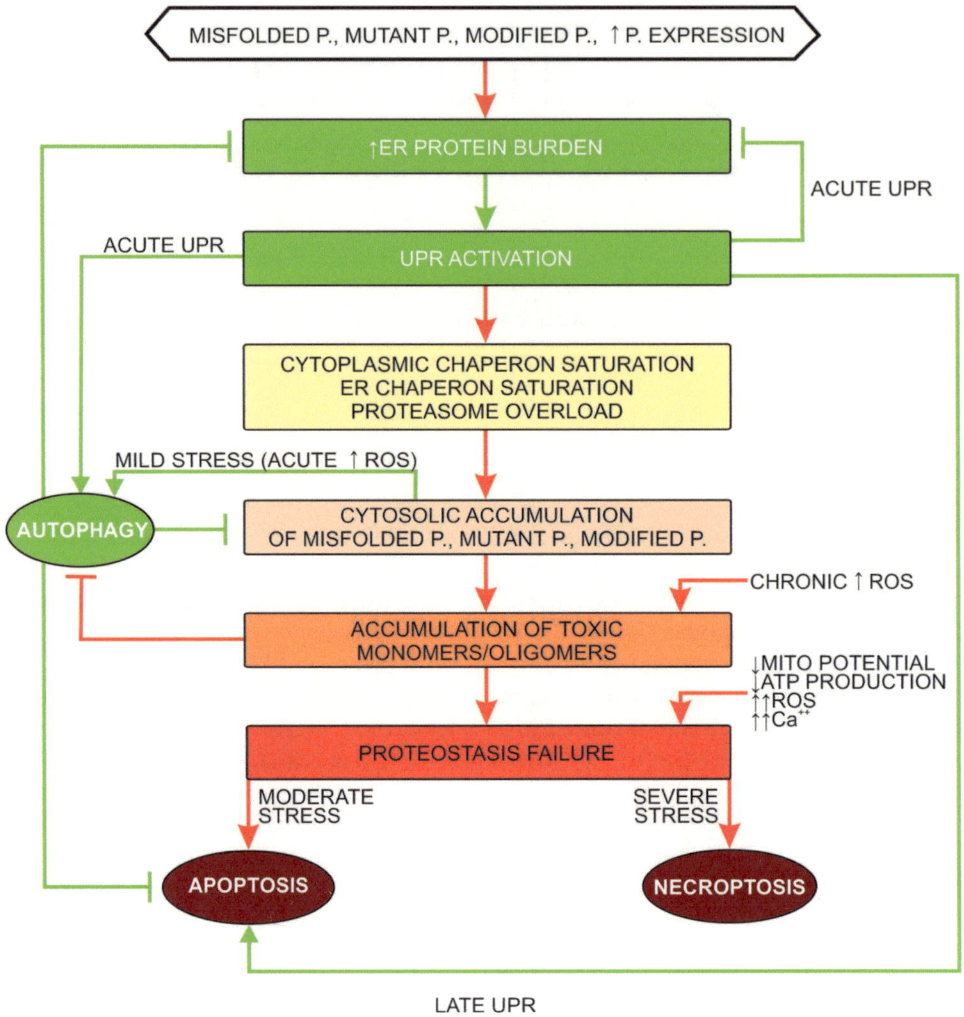

Figure 1.
Proteostasis in human neurodegenerative brain disorders (NBD). Abbreviations: P (proteins), ER (endoplasmic reticulum), UPR (unfolded protein response), ROS (reactive oxygen species); red lines and arrows indicate progressive failure of proteostasis ultimately leading to NBD. Green arrows and lines indicate appropriate responses of proteostasis to altered P that prevent or slow down the progress of NBD.

oligomers; their production is stimulated by chronic inflammation and increased reactive oxygen species (ROS) production. These monomers and oligomers repress AUT and trigger either apoptosis or necroptosis (**Figure 1**) [4, 6–8].

1.2 Autophagy changes in selected NBD

An efficient autophagy (AUT) delays or attenuates the progression of AD, PD and PrD [9–12]. A summary of AUT changes in selected NBD is shown in **Figure 2**. Post-translationally modified proteins (PTMP)—such as soluble amyloid β-peptide 42 with a single oxidised methionine residue at position 35 (Aβ42-MET35-OX) in Alzheimer's disease, alpha-synuclein oxidised on methionine residues (MET-OX-αSYN) in Parkinson's disease and oxidised, self-propagating infectious isoforms of prion protein (MET-OX-PRP^{Sc}) in prion diseases (PrD)—inhibit (a) AUT, in AD, PD and PrD, and also (b) mitochondrial (MITO) function [13–23]. MET-OX-PRP^{Sc} indirectly damage MITO function. The normal prion protein (PrP^{c}) binds with

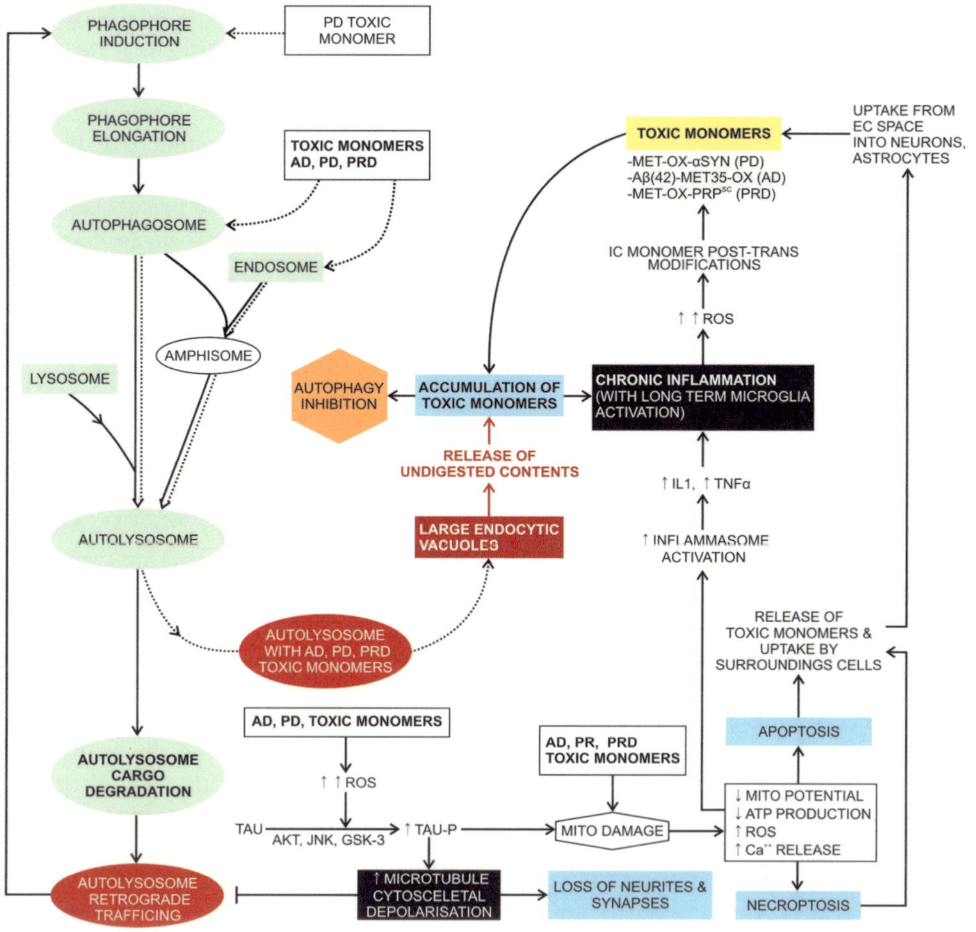

Figure 2.
Summary of AUT changes in selected NBD. Abbreviations: AKT (protein kinase B), GSK3 (glycogen synthase kinase 3), JNK (c-Jun N-terminal kinase), TAU (TAU protein), TAU-P (phosphorylated TAU protein).

a variety of molecules, including copper ions [24, 25], and PrP^c expression levels correlate with Cu/Zn superoxide dismutase, glutathione reductase and cytochrome c oxidase activities [26]. These observations support the hypothesis that PrP^c is (a) an important endogenous scavenger, protecting structural and signalling proteins from oxidation, due to its high number of methionine residues, and (b) vital for the intracellular transport of copper to superoxide dismutase, which is dependent on copper binding for its antioxidant function. Loss of PrP^c, due to conversion to PrP^{Sc} and MET-OX-PrP^{Sc}, which do not bind copper and have a reduced antioxidant activity, reduces the cell's intracellular antioxidant and copper transport capacity and precipitates MITO dysfunction, due to an increased oxidation of cytochrome c oxidase and other MITO proteins [27–30].

AUT is inhibited at the stage of protein digestion (during autolysosome cargo degradation) by the undigestible PTMP and is diverted to the formation of large endocytic vacuoles that rupture and release the undigested PTMP into the cytosol, thus progressively increasing their intracellular concentration. PTMP of AD and PD accelerate microtube cytoskeletal depolarisation, thus blocking autolysosome retrograde trafficking and accelerating loss of neurites, synapses and synaptic transmission [31–39]. PTMP inhibition of MITO function leads to (a) a reduced ATP production and an increased MITO release of ROS and Ca^{2+} into the cytosol [38, 40–44] and

(b) activation of inflammasomes with an increased release of cytokines interleukin 1 (IL1), from microglia, and tumour necrosis factor alpha (TNFα), from astrocytes and neurons, and finally apoptosis or necroptosis [38, 45–52]. Apoptosis or necroptosis of nerve cells and astrocytes releases PTMP and their oligomers into the extracellular space, thus contributing to the spread of inflammation and neurodegenerative disorder in the brain. The physiological process of apoptosis that normally prevents the spill of cell's molecules to the extracellular space is perturbed by the altered proteostasis into a pathological one in NBD. This transformation is sustained by several intracellular processes including the accumulation of undigestible PTMP, increased oxidative stress, and distorted expression of apoptotic proteins [53–56].

The AUT capacity of brain cells is important in the regulation of immune responses and inflammation that occur in NBD [57, 58]. Protein aggregates (aggresomes), present in age-related NBD, activate inflammasomes. Activated inflammasomes lead to a low-grade inflammation associated with a declined autophagic capacity [59]. On the other hand, autophagy attenuation leads to inflammasome precipitated excessive caspase-1 activation and elevated IL-1β secretion in response to lipopolysaccharide (LPS) stimulation [10, 60, 61]. Also, ER stress and inflammation coexist in NBD, for example, in AD, and are intertwined [57]. Chronic neuroinflammation (CNI) develops into a self-damaging process and is an important factor in sustaining NBD including AD, PD and PRD. CNI includes activation of microglia and astrocytes and infiltration of peripheral immune cells. Transient activation of microglia, accompanied by the release of inflammatory cytokines that amplify the inflammatory response by activating and recruiting astrocytes and peripheral immune cells to the brain lesion, ensures the brain's integrity by removing foreign bodies and cell debris. CNI is toxic to neurons due to sustained release of inflammatory cytokines (e.g. ILs 1β and 6, TNFα) and ROS and microglial phagocytosis of neighbouring intact nerve cells, thus contributing to the development and progression of NBD. The progressive loss of neurons further contributes to generation of cell debris and sustains microglial hyperactivation [62].

The detrimental effects of PTMP, sustained inflammation and increased ROS production are further exacerbated by the formation of AUT-resistant soluble Aβ oligomers (AβO) in AD and AUT-resistant αSYN oligomers in PD that further stimulate chronic inflammation and increased cytosolic ROS, contributing to apoptosis or necroptosis of neurons. Therefore, activation of apoptosis or necroptosis in AD, PD or PrD is triggered by a positive feedback loop between chronic inflammation in the brain (to which astrocytes and microglia are the main contributor) and the production of PTMP. In addition to high levels of ROS, the production of PTMP in the cytosol is facilitated by copper ions in AD [63] and by iron ions, dopamine and accumulation of alpha-synuclein (the precursor of oxidised αSYN monomer) in PD [17]. Although chronic brain inflammation contributes to the process of PrP^{Sc} production, it is not necessary to sustain it, since the PrP^{Sc} only needs the PrP^{c} molecules for its propagation [64].

2. Crosstalk among AUT, apoptosis and necroptosis signalling pathways in selected NBD

AUT, apoptosis and necroptosis have interlinked signalling pathways. Examples of key signalling molecules that regulate the transition among these three processes are presented in Section 2.1. The crosstalk among AUT, apoptosis and necroptosis signalling pathways, with the potential sites of modulation by Alzheimer's, Parkinson's and prion diseases (PrD), is summarised in **Figure 3**.

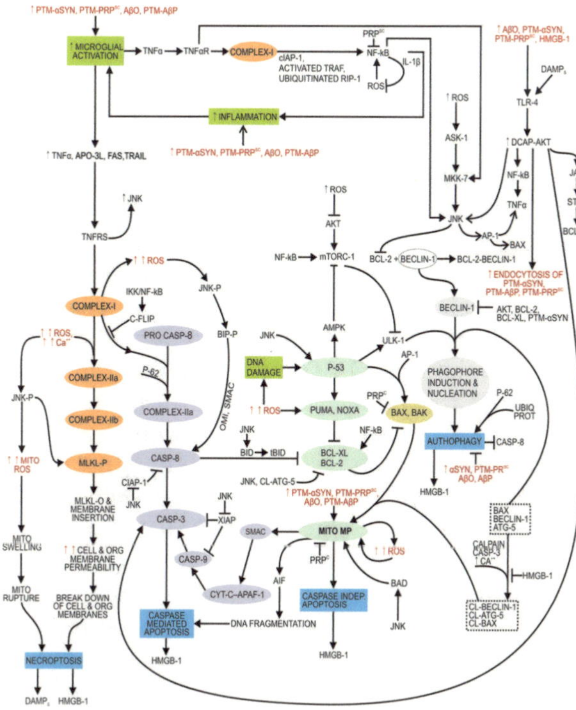

Figure 3.
Crosstalk among AUT, apoptosis and necroptosis signalling pathways with the potential sites of modulation by AD, PD and PrD. Abbreviations: αSYN (alpha-synuclein); AβO (amyloid β oligomers); AβP (amyloid β monomers with 39 to 42 amino acid residues); AIF (apoptosis-inducing factor); AKT (protein kinase B); AMPK (5' AMP-activated protein kinase); AP-1 (activator protein 1); APAF-1 (apoptotic protease activating factor 1); APO-3L (APO3 ligand); ASK-1 (apoptosis signal-regulating kinase 1); ATG-5 (AUT-related 5); BAD (Bcl2-associated agonist of cell death); BAK (Bcl-2 homologous antagonist/killer); BAX (apoptosis regulator BAX); BCL-2 (B-cell lymphoma 2); BCL-XL (B-cell lymphoma-extra large); Beclin-1 (mammalian ortholog of the yeast AUT-related gene 6 (ATG-6)); BID (BH3 interacting domain death agonist); BIP-P (phosphorylated binding immunoglobulin protein); C-FLIP (FADD-like IL-1β-converting enzyme-inhibitory protein); calpain (proteolytic enzyme, a protein belonging to the family of calcium-dependent, non-lysosomal cysteine proteases); CASP-3, CASP-8/10, CASP-9 (caspase-3, caspase-8/10, caspase-9); cIAP-1 (cellular inhibitor of apoptosis protein 1); CL-ATG5 (cleaved AUT-related 5 (ATG5) protein); CL-BAX (cleaved apoptosis regulator BAX); CL-Beclin-1 (cleaved mammalian ortholog of the yeast AUT-related gene 6); Complex-I (TNFα bound to TNFα receptor that is associated with TRADD (tumour necrosis factor receptor type 1-associated death domain protein), RIPK1 (receptor-interacting serine/threonine-protein kinase 1), TRAF2 (TNF receptor-associated factor 2) and cIAP-1/2 (cellular inhibitor of apoptosis protein 1 and 2)); Complex-IIa (pro-caspase-8, RIPK1, FADD (FAS-associated protein with death domain)); Complex-IIb (pro-caspase-8, RIPK1, RIPK3 (receptor-interacting serine/threonine-protein kinase 3), FADD, MLKL (mixed lineage kinase domain-like pseudokinase)); CYT-C (cytochrome c); DAMPs (damage-associated molecular patterns); DCAP-AKT (activation of toll/IL-1R (TIR) domain-containing adaptor proteins (e.g. mal, TRIF, TRIF-related adaptor molecule, IL-1R-associated kinase-1, IL-1R-associated kinase-M, MAPK, TNFR-associated factor 6, toll-interacting protein)); FAS (apoptosis antigen 1); HMGB-1 (high-mobility group box 1 protein); IKK (IκB kinase enzyme complex, part of the upstream NF-κB signal transduction cascade); IL-1β (interleukin-1 beta); JAK (Janus kinase); JNK (c-Jun N-terminal kinase); JNK-P (phosphorylated c-Jun N-terminal kinase); MITO (mitochondrial); MITO MP (mitochondrial membrane permeability); MITO ROS (mitochondrial reactive oxygen species); MKK7 (MAP kinase kinase 7); MLKL-O (MLKL oligomerisation with translocation and insertion into cell's and organelles' membranes with increased permeability); MLKL-P (phosphorylated pseudokinase mixed lineage kinase domain-like protein); mTORC1 (mammalian target of rapamycin complex 1); NF-κB (nuclear factor kappa-light-chain-enhancer of activated B cells protein complex); NOXA (adult T cell leukaemia-derived PMA-responsive); OMI alias HtrA2 (serine protease HTRA2); ORG (cell organelles); p53 (cellular tumour antigen p53); p62 (nucleoporin p62 protein complex associated with the nuclear envelope); PRO CASP 8/10 (pro-caspase-8/10); PRP^{C} (normal form of prion protein); PRP^{SC} (self-propagating, protease-resistant, infectious isoforms of prion protein); PTM (post-translationally modified); PUMA (p53 upregulated modulator of apoptosis); RIP-1 (receptor-interacting serine/threonine-protein kinase 1); ROS (reactive oxygen species, e.g. peroxides, superoxide, hydroxyl radical or singlet oxygen); SMAC (second mitochondria-derived activator of caspases); STAT (signal transducer and activator of transcription 3/5); tBID (truncated BID protein); TLR-4 (toll-like receptor 4, member of the pattern recognition receptor (PRR) family); TNFα (tumour necrosis factor alpha); TNFαR (tumour necrosis factor alpha receptor); TNFRS (tumour necrosis factor receptor superfamily); TRAF (TNF receptor-associated factor); TRAIL (TNF-related apoptosis-inducing ligand); UBIQ PROT (ubiquitinated proteins); ULK1 (serine/threonine-protein kinase ULK1); XIAP (X-linked inhibitor of apoptosis protein).

2.1 Examples of signalling molecules that regulate crosstalk among AUT, apoptosis and necroptosis pathways in selected NBD

Intracellular *adenosine triphosphate* (ATP) promotes either apoptosis or necroptosis in a concentration-dependent manner; high ATP levels promote apoptosis, and low ATP levels promote necroptosis [65, 66]. Therefore, ATP production in the MITO determines the type of cell death. The best understood inflammation- and necroptosis-promoting cytokine that modulates mitochondrial ATP and ROS levels is TNFα [67]. As explained above, PTMP inhibition of MITO function leads to activation of inflammasomes with an increased release of tumour necrosis factor alpha (TNFα) from astrocytes and neurons [38, 45–52]. The sustained TNFα stimulation in NBD is the result of two mechanisms. (a) The PTMP of AD, PD and PrD are not digested by AUT; they accumulate in affected cells by their release into the cytosol from endolysosomal and autolysosomal compartment together with proteolytic enzymes [68]. (b) The PTMP in PD and PrD spread through the brain by a prion-like mechanism [69, 70]. The sustained TNFα stimulation can lead to over-activation of PARP1, a nuclear DNA repair enzyme that is activated by DNA damage, due to an increased MITO ROS production. PARP1 over-activation precipitates an acute depletion of NAD^+, inhibition of oxidative phosphorylation with a severe drop in ATP production and a subsequent activation of necroptosis [65, 71–73].

AUT-related 5 (Atg5) protein stimulates elongation of autophagosome membranes that envelope PTMP into autophagosomes [74–76] and also regulates the balance between AUT and apoptosis [77]. The neurons' cytosolic Ca^{2+} is increased in NBD due to the PTMP elicited (a) ER and MITO release of Ca^2 into the cytosol [4–7, 78] and (b) an increased Ca^{2+} entry through the N-methyl-D-aspartate (NMDAR) glutamate receptor and ion channel proteins from the extracellular space [79–81]. Increased cytosolic Ca^{2+} promotes calpain-1- and calpain-2-mediated cleavage of ATG5, with a loss of pro-AUT function and concomitant triggering of cytochrome c-/caspase-mediated apoptosis due to the inhibition of Bcl-xL in the MITO by the cleaved ATG5 [82]. The calpain-1- and calpain-2-mediated cleavage of ATG5 is attenuated by decreased levels of cytosolic Ca^{2+} [83]. Cytosolic HMGB1 attenuates apoptosis by protecting the AUT proteins beclin 1 and ATG5 from calpain-mediated cleavage during inflammation [84].

Beclin 1 stimulates AUT [16, 85, 86]; an enhanced AUT has a concomitant anti-apoptotic effect by clearing apoptosis-associated molecules, for example, active caspase-8 [87–89]. Beclin 1 is cleaved by caspases, thus losing its pro-AUT function, and the cleaved beclin 1 (i.e. C-terminal beclin 1 fragment) promotes apoptosis by triggering the release of MITO cytochrome c [90–92].

B-cell lymphoma 2 (Bcl-2) family of proteins regulate MITO apoptotic pathway and also AUT; for example, Bcl-2 and Bcl-xL inhibit AUT and apoptosis [93, 94]. Bcl-2 and Bcl-xL proteins have an anti-apoptotic effect, whereas Bax, Bad, Bid, Bim, Bmf, PUMA and NOXA promote apoptosis. Calpain-mediated cleavage of Bax, induced by high cytosolic Ca2+, mediates apoptosis [82]. The interactions between anti-apoptotic and pro-apoptotic Bcl-2 family members determine the activation of apoptosis [95–104]. Bcl-2 and Bcl-xL associate with beclin 1 and supress the beclin 1-dependent autophagic activation [105]. This AUT suppression can be abolished by the pro-apoptotic Bcl-2 family proteins (e.g. Bad, Bid) [106]. The inhibition of Bcl-2 on beclin 1 is also attenuated by phosphorylation of Bcl-2 by JNK-1 or Beclin-1 by DAPK1, thus promoting AUT [107, 108]. Increased expression of Bak, Bad, Bcl-2 and Bcl-x was observed in AD [109]. Cytosolic PrPc protects human primary neurons from Bax-mediated apoptosis [110–112]; therefore the PrP^{sc}-precipitated reduction should facilitate apoptosis in PrD.

Caspase-8 activity is changed in NBD. It has been suggested that an increased caspase-8 activity, associated with an increased caspase-3 activity in the same hippocampal tissue sections from patients with AD, contributes to the development of AD in humans. Recently, two caspase-8 variants, with a reduced activity and associated with an increased risk for development of AD in human, were identified. This finding is consistent with the multiple AD-related changes in the human brain, including loss of synaptic plasticity and memory function and increased microglia pro-inflammatory activation [113]. Caspase-8, within the death-inducing complex II, triggers either apoptotic or necroptotic cell death. Activated caspase-8 promotes apoptosis and also inhibits necroptosis by cleaving RIPK1, RIPK3 and CYLD [114–116], thus preventing CYLD-mediated deubiquitylation of RIPK1 and subsequent RIPK1 kinase activation and necroptosis [117]. The association of caspase-8 with pseudo-caspase cFLIP suppresses apoptosis and also necroptosis, since the residual levels of caspase-8 activity are still sufficient to cleave and inactivate RIPK1 and RIPK3 [118].

c-Jun N-terminal kinase (JNK) promotes either apoptosis or necroptosis, depending on its upstream signalling pathways. JNK is required for apoptosis of central nervous system neurons [119]. JNK promotes apoptosis by several signalling pathways that were characterised in different cell experimental models. It is unlikely that all of the observed JNK's pro-apoptotic effects are present in all of the cells at the same time [120]. However, it is important to be aware of the JNK's ability to modulate apoptosis at different levels. To summarise, the known pro-apoptotic effects of JNK are: (a) Activated MAP2Ks phosphorylate JNK and phosphorylated JNK translocates to the nucleus and phosphorylates c-Jun [121, 122] that promotes AP-1 expression; AP-1 promotes transcription of pro-apoptotic proteins TNF-α, Fas-L and Bak [123–125]. (b) JNK phosphorylates p53, enhancing the expression of pro-apoptotic genes Bax and PUMA [126–128]; the increased Bax expression and translocation to mitochondria is sufficient to promote MITO outer membrane permeabilization, the consequent release of cytochrome c and caspase-9 and caspase-3 activation [129–133]. (c) JNK phosphorylates 14-3-3-associated Bad, thus promoting its translocation into MITO and subsequent release of cytochrome c [134, 135]. (d) JNK phosphorylates pro-apoptotic proteins Bim and Bmf, and these phosphorylated proteins activate Bax and/or Bak [136–140]. (e) Phosphorylated Bim binds to and inhibits the Bcl2's anti-apoptotic activity, thus increasing the probability of MITO-activated apoptosis [141, 142]. (f) JNK inhibits the anti-apoptotic Bcl2 by phosphorylation, to induce apoptosis [143, 144]. (g) JNK has the ability to promote apoptosis by stimulating the activity of many pro-apoptotic signalling molecules. (h) Activation of TNFRS (e.g. TNFR1, DR3-6) can lead to apoptosis [144, 145]. (i) Activation of DRs and TNFα receptors stimulates JNK activation that promotes apoptosis by increased expression of DRs [146, 147]; increased expression of pro-apoptotic proteins Bak, Bim and Bax [148, 149]; inhibition of anti-apoptotic proteins XIAP (caspase-3, caspase-7 and caspase-9 inhibitor) and cIAP1 (caspase-8 inhibitor) [150, 151]. JNK's role in NBD is best understood in AD; JNK activation is positively correlated with AD progression [152]. Amyloid-β protein fragments activate JNK [153, 154]. Also, JNK phosphorylates tau, thus promoting (a) microtubule cytoskeleton breakdown, (b) attenuation of intracellular transport and (c) loss of synaptic terminals [155–166].

Activation of tumour necrosis factor receptor superfamily (e.g. TLRs or TNFαR) or DNA damage can trigger necroptosis by activation of the Complex I-IIa-IIb-phosphorylated pseudokinase mixed lineage kinase domain-like protein (MLKL) signalling pathway; the final steps are (a) RIP3-dependent phosphorylation of MITO proteins PGAM5 and Drp-1 (increasing MITO ROS production); (b) insertion of phosphorylated MLKL into the MITO membrane with the cumulative effects of increased MITO membrane permeability, loss of membrane potential, decreased ATP and increased ROS production [120, 167–169]; and (c) phosphorylated MLKL's translocation to the plasma membrane and activation of Ca^{2+} influx through plasma membrane channels with concomitant plasma membrane breakdown [169]. Increased cytosolic ROS production inactivates MAP kinase phosphatase 1, enabling sustained activation of phosphorylated JNK; phosphorylated JNK promotes necroptosis by (a) stimulating MLKL phosphorylation and by (b) promoting cytochrome c release from MITO via activation of BID [170, 171].

FLICE inhibitory proteins (FLIPs). Under stress-free conditions, FLIPs (FLICE inhibitory proteins) attenuate LC3's binding with ATG3, thus preventing ATG3-mediated elongation of autophagosomes and AUT. During stress, FLIPs allow for ATG3-LC3 interaction and stimulate AUT. Therefore, FLIPs (e.g. C-FLIP) can inhibit apoptosis and also AUT [172].

The *high-mobility group box protein 1* (HMGB1) is a nuclear protein released by glia and necrotic or hyper-excitatory neurons after inflammasome activation; it activates receptors for advanced glycation end products (RAGE) and the toll-like receptor (TLR) 4 on neurons and microglia [173, 174]. When HMGB1 binds to TLR4 on neurons, it phosphorylates MARCKS via MAP kinases and induces neurite degeneration, present in AD [173]. The disulphide form of HMGB1 potentiates the microglia pro-inflammatory response; therefore, repeated releases of HMGB from damaged nerve cells during chronic neuroinflammation in PD and AD could lead to an exacerbated neuroinflammatory response of microglia [175–177]. HMGB1, in a rat model of AD, caused (a) inhibition of microglial amyloid β-peptide 42 clearance and enhanced amyloid β-peptide 42 neurotoxicity [178] and (b) dysfunction of microglial amyloid β-peptide 40 phagocytosis [179].

The *nuclear factor kappa-light-chain-enhancer of activated B cells protein complex* (NF-κB) signalling pathway was repressed in a prion-infected cell line and animal brain tissues as evidenced by a decreased level of transcription factor p65/nuclear factor NF-kappa-B p65 subunit (p65) and downregulation of phosphoinositide 3-kinase (PI3K) and protein kinase B (PKB/Akt) in both experimental models [180]. In AD cell models, the exposure to amyloid β-peptide or amyloid precursor protein induced NF-κB activation [181, 182], and inhibition of NF-κB transcriptional activity increased neuronal death in the presence of amyloid β-peptide [183]. NF-κB activation can protect neurons against amyloid β-peptide-induced cell death [184]. Patients with PD have an increased percentage of dopaminergic neurons in the substantia nigra with nuclear p65 immunoreactivity [185]. NF-κB is one of the several factors that regulate Beclin-1 expression; Beclin-1 promotes AUT by stimulating autophagosome formation [186–188]. Increased NF-κB activation in the brain, in addition to stimulating AUT, protects nerve cells against NBDs' mediated injury by several mechanisms including increased transcription of MITO antioxidant enzyme manganese superoxide dismutase (MnSOD) and Bcl-xL genes [189].

Sirtuins (SIRTs), NAD^+-dependent protein deacetylases, modulate apoptosis and necroptosis [190]. For example, SIRT1 promotes AUT by deacetylation of ATG5, ATG7 and ATG8 [191]. Following TNFα receptor stimulation, SIRT2 promotes the association of RIP1 and RIP3, the subsequent formation of complex II and necroptosis [192]. In animal and cell culture models of AD, SIRT1 reduces neurodegeneration in mouse hippocampus and promotes primary neuronal survival [193]. The reduced SIRT1 mRNA and protein levels are associated with an accumulation of amyloid β-peptide 42 and tau in the brains of AD patients [194].

Tumour protein p53 (p53) modulates AUT and apoptosis. It promotes apoptosis by Bax activation in the cytoplasm; BAX initiates apoptosis by triggering mitochondrial cyt c release and caspase-3 activation [195]. In the nucleus, p53 activates transcription of Bax, PUMA and Noxa [196]. PUMA displaces cytoplasmic p53 from the Bcl-xL-p53 complex, promoting p53 activation of the apoptotic pathway [197]. In the nucleus, p53 also stimulates AUT through transcription activation of ULK1, sestrin1/2 and damage-regulated AUT modulator (DRAM) [198, 199]. Indirectly, p53 promotes AUT by mTOR inhibition, via activation of AMP-dependent kinase and tuberous sclerosis (TSC) 1/TSC2 complex pathway [200]. It was suggested that DRAM has a dual role of promoting either AUT- or p53-mediated apoptosis [201]. In a *Drosophila* model of AD tauopathy, p53 prevented neurodegeneration by increased expression of amphiphysin, clathrin light chain, clathrin heavy chain, RAS oncogene family and synaptotagmin β synaptic genes [202]. p53 levels are significantly increased in brains of patients with AD [203] and are correlated with brain MITO dysfunction [204]. Recently, it was suggested that tau oligomers sequester and downregulate functional phospho-p53 in an AD mouse model and in patients with AD [205].

Ubiquitin-binding protein p62 (p62) modulates cell death switching between apoptosis and necroptosis. In a cell model, p62 promotes either necroptosis, when p62 is associated with the necrosome (i.e. complex II), or apoptosis when the P62-necrosome association is blocked [206]. The p62 regulates apoptotic and autophagic processes [207]. P62 mediates AUT degradation by first binding polyubiquitinated proteins with the ubiquitin-associated domain and then to autophagosomes through the LC3-interacting region [208, 209]. In response to tumour necrosis factor receptor stimulation, P62 promotes apoptosis by stimulating activation of caspase-8 [210, 211]. The levels of p62 are increased in NBD, for example, in PrD [212, 213]. Autophagy disposal of aberrant proteins is stimulated by the p62-Keap1-NRF2 signalling pathway [214]. For example, in a mouse model of AD, increased brain p62 expression improved cognition by an autophagy-mediated mechanism that reduced amyloid β-peptide 40/42 levels [215].

2.2 Summary of similarities/differences in the mechanistic pathways between selected NBD

Beclin-1, ATG-5, NF-κB, JNK, p53, p62, HMGB1 and ROS are the key signalling molecules that mediate crosstalk among AUT, apoptosis and necroptosis. ATG5 and Beclin-1 in conjunction with ULK-1 and BAX promote AUT by initiating phagophore induction and nucleation steps. Cleavage of ATG-5 and Beclin-1 by calpain, caspase-3 or increased cytosolic free calcium changes their function from stimulating AUT to promoting apoptosis via increased MITO membrane permeability. Cleaved ATG5 inhibits the anti-apoptotic activity of BCL2 and BCL-XL on BAX and BAK, further promoting increased MITO membrane permeability and apoptosis. P53 activation plays a dual role by promoting apoptosis (via activation of PUMA and NOXA) and AUT by ULK1 activation. The JNK signalling kinase blocks the binding of BCL-2 to Beclin-1, thus enabling Beclin-1 to participate in AUT initiation, and also activates the apoptosis-triggering proteins BAX and BAK. Phosphorylated JNK promotes necroptosis by stimulating MLKL phosphorylation and apoptosis by caspase-8 activation. p62 promotes AUT and apoptosis. HMGB-1 is released during AUT, apoptosis and necroptosis, and by inhibiting the cleavage of ATG-5, BAX and Beclin-1 simultaneously promote AUT and inhibit apoptosis. Mild increases in cytosolic ROS act as signalling molecules that promote a physiological balance between AUT, apoptosis and necroptosis, which favour AUT; moderate and high increases in cytosolic ROS concentrations favour apoptosis and necroptosis over AUT. The products of post-translational protein modifications in AD, PD and PrD favour apoptosis and necroptosis over AUT by (a) increasing the activation of apoptosis (e.g. by increasing MITO membrane permeability) and necroptosis, by chronic activation of TRL4 and TNFα receptors [216–234],

(b) promoting moderate to high increases in cytosolic ROS concentrations and (c) attenuating AUT [42, 62, 235–240]. In contrast to PD and AD, PrP^{SC}-infected cells are more likely to respond with necroptosis and then apoptosis. For example, a significant upregulation of necroptosis signalling molecules phosphorylated MLKL, MLKL and receptor-interacting serine/threonine-protein kinase 3 (RIP3) was measured in the post-mortem cortical brains of patients with various types of human PRD [241].

3. Pharmacological strategies targeting AUT, apoptosis and necroptosis signalling pathways

At present, most of the studies, devoted to the development of pharmacological interventions for NBD, are focused on the crosstalk of AUT and apoptosis signalling pathways in neurons. Future research should also include development of pharmacological interventions that target other cells involved in the development of NBD, including microglia, astrocytes, endothelial cells and pericytes [242]. The development of pharmacological interventions for NBD should be guided by several key questions: (a) How to modulate the role of AUT from pro-death to pro-survival? (b) How is the information from the crosstalk among AUT, apoptosis and necroptosis integrated? (c) How to modulate the crosstalk among AUT, apoptosis and necroptosis? and (d) How is the information from the crosstalk among AUT, apoptosis and necroptosis (e.g. inflammation-promoting molecules) shared among different cells involved in the development of NBD? [242]. Examples of pharmacological strategies are given below:

Pharmacological strategies to ameliorate MITO dysfunction include:

(a) Targeting excessive ROS production:

- (a1) Mercaptamine that increases levels of glutathione in human [78].
- (a2) Antioxidant vatiquinone used in clinical trials [243].
- (a3) RTA-308 stimulates Nrf2 to enhance the expression of pro-oxidant genes and to repress inflammatory genes in an animal model [244].
- (a4) Antioxidants coenzyme Q, lipoic acid and green tea polyphenol epigallocatechin gallate attenuate the effects of NBD in animal models [245–248].
- (a5) Ceria nanoparticles are ROS scavengers that localise in MITO and suppress neuronal death in an AD mouse model [249].

(b) Targeting mitochondrial biogenesis: stimulation of PGC1-α's ROS scavenging activity with SIRT1 could attenuate ROS-induced damage in AD [250].

AUT inducers are (a) mTOR inhibitors, either ATP-competitive inhibitors (e.g. Torin1 and related compounds) or non-ATP-competitive inhibitors (e.g., rapamycin and rapalogs), and (b) acting by mTOR-independent targets [238]. The most promising AUT inducers, acting by mTOR inhibition, are the non-ATP-competitive inhibitors rapamycin and rapalogs that are mTORC1 selective and induced AUT in animal models of AD, PD and PrD [251–258]. The AMPK signalling pathway is activated by mTOR-independent AUT activators, for example, by trehalose. Trehalose inhibits GLUT proteins, thus eliciting AMPK activation [259]. Trehalose-induced

AUT induction, with concomitant therapeutic effects, was demonstrated in mouse models of NBD, including AD, PD and PrD [260–265].

TNFα signalling pathway is the focus of pharmacological interventions targeting neuroinflammation in NBD with a variety of compounds [57]: (a) serotonin binds to microglial receptors and has anti-inflammatory effects; serotonin treatment reduced TNFα release in cultured primary microglia cells exposed to AβO and in mouse brains infused with AβO and also prevented AD-associated behavioural changes [266]; (b) etanercept, a decoy TNF receptor and IgG1 Fc fusion protein that inhibits the binding of soluble TNF to cell-surface TNF receptors, was evaluated in several clinical trials on patients with AD; no statistically significant results were reported; however, the drug was well tolerated, and large-scale trials are expected [57]; and (c) infliximab, a human monoclonal antibody that binds TNFα and was used to treat human auto-immune and inflammatory diseases, prevented eIF2a phosphorylation and long-term memory loss in a mouse model of AD [7, 267].

4. Conclusions

Neurodegenerative brain disorders (NBD) change brain cell proteostasis due to the accumulation of normal, mutant, misfolded or unfolded proteins in the endoplasmic reticulum (ER). The increased ER burden elicits the unfolded protein response (UPR) and stimulates AUT. In the short term, these responses tend to attenuate ER's stress, by reducing the ER's protein load and increasing the ER's folding capacity. In the long term, with prolonged ER stress, the UPR changes from supporting cell survival to promoting apoptosis. The failure of the ER stress response to meet the increased protein burden is reflected in an increased cytosolic protein accumulation that initially further stimulates AUT. Over time, the accumulated proteins in the cytosol undergo post-translational changes into toxic monomers and oligomers that repress AUT at multiple levels and promote either apoptosis or necroptosis. Apoptosis and necroptosis of the affected cells lead to the release of toxic proteins into the surrounding tissue and trigger the response of microglia and astrocytes. Chronic neuroinflammation, sustained by the spread of progressive failure of AUT among brain cells, due to the release of toxic monomers and oligomers from dying cells and their uptake by initially healthy cells and by the persistent activation of microglia and astrocytes by toxic monomers and oligomers, also contributes to nerve apoptosis or necroptosis. The signalling pathways of apoptosis, AUT and necroptosis are interlinked. A better understanding on how chronic neuroinflammation, Alzheimer's, Parkinson's and prion diseases modulate the crosstalk among these signalling pathways could contribute to the development of new therapeutic interventions for these NBD.

Acknowledgements

The author thanks Professor Irina Milisav for reviewing the manuscript and suggesting improvements. The assistance of Ms. Vanja Mavrin in drawing the final figures is acknowledged.

Conflicts of interest

The author declares no conflict of interest.

Funding

This work was supported by ARRS grant number P3-0171.

Author details

Samo Ribarič* and Irina Milisav Ribarič
Faculty of Medicine, Institute of Pathophysiology, University of Ljubljana, Ljubljana, Slovenia

*Address all correspondence to: samo.ribaric@mf.uni-lj.si

References

[1] Serrano-Pozo A, Frosch MP, Masliah E, Hyman BT. Neuropathological alterations in Alzheimer disease. Cold Spring Harbor Perspectives in Medicine. 2011;**1**(1):a006189. DOI: 10.1101/cshperspect.a006189

[2] Dickson DW. Parkinson's disease and parkinsonism: Neuropathology. Cold Spring Harbor Perspectives in Medicine. 2012;**2**(8):1-15. DOI: 10.1101/cshperspect.a009258

[3] Soto C, Satani N. The intricate mechanisms of neurodegeneration in prion diseases. Trends in Molecular Medicine. 2011;**17**(1):14-24. DOI: 10.1016/j.molmed.2010.09.001

[4] Martin-Jimenez CA, Garcia-Vega A, Cabezas R, Aliev G, Echeverria V, Gonzalez J, et al. Astrocytes and endoplasmic reticulum stress: A bridge between obesity and neurodegenerative diseases. Progress in Neurobiology. 2017;**158**:45-68. DOI: 10.1016/j.pneurobio.2017.08.001

[5] Gerakis Y, Hetz C. A decay of the adaptive capacity of the unfolded protein response exacerbates Alzheimer's disease. Neurobiology of Aging. 2018;**63**:162-164. DOI: 10.1016/j.neurobiolaging.2017.09.012

[6] Gerakis Y, Hetz C. Emerging roles of ER stress in the etiology and pathogenesis of Alzheimer's disease. The FEBS Journal. 2018;**285**(6):995-1011. DOI: 10.1111/febs.14332

[7] Freeman OJ, Mallucci GR. The UPR and synaptic dysfunction in neurodegeneration. Brain Research. 2016;**1648**(Pt B):530-537. DOI: 10.1016/j.brainres.2016.03.029

[8] Shah SZA, Zhao D, Hussain T, Yang L. The role of unfolded protein response and mitogen-activated protein kinase signaling in neurodegenerative diseases with special focus on prion diseases. Frontiers in Aging Neuroscience. 2017;**9**:120. DOI: 10.3389/fnagi.2017.00120

[9] Bingol B. Autophagy and lysosomal pathways in nervous system disorders. Molecular and Cellular Neurosciences. 2018;**91**:167-208. DOI: 10.1016/j.mcn.2018.04.009

[10] Lai M, Yao H, Shah SZA, Wu W, Wang D, Zhao Y, et al. The NLRP3-caspase 1 inflammasome negatively regulates autophagy via TLR4-TRIF in prion peptide-infected microglia. Frontiers in Aging Neuroscience. 2018;**10**:116. DOI: 10.3389/fnagi.2018.00116

[11] Fan XY, Tian C, Wang H, Xu Y, Ren K, Zhang BY, et al. Activation of the AMPK-ULK1 pathway plays an important role in autophagy during prion infection. Scientific Reports. 2015;**5**:14728. DOI: 10.1038/srep14728

[12] Wang H, Tian C, Sun J, Chen LN, Lv Y, Yang XD, et al. Overexpression of PLK3 mediates the degradation of abnormal prion proteins dependent on chaperone-mediated autophagy. Molecular Neurobiology. 2017;**54**(6):4401-4413. DOI: 10.1007/s12035-016-9985-0

[13] Grant CM. Sup35 methionine oxidation is a trigger for de novo [PSI(+)] prion formation. Prion. 2015;**9**(4):257-265. DOI: 10.1080/19336896.2015.1065372

[14] Nixon RA. Autophagy, amyloidogenesis and Alzheimer disease. Journal of Cell Science. 2007;**120**(Pt 23):4081-4091. DOI: 10.1242/jcs.019265

[15] Pickford F, Masliah E, Britschgi M, Lucin K, Narasimhan R, Jaeger PA, et al. The autophagy-related protein beclin 1 shows reduced expression in early Alzheimer disease and regulates

amyloid beta accumulation in mice. The Journal of Clinical Investigation. 2008;**118**(6):2190-2199. DOI: 10.1172/JCI33585

[16] Salminen A, Kaarniranta K, Kauppinen A, Ojala J, Haapasalo A, Soininen H, et al. Impaired autophagy and APP processing in Alzheimer's disease: The potential role of Beclin 1 interactome. Progress in Neurobiology. 2013;**106-107**:33-54. DOI: 10.1016/j.pneurobio.2013.06.002

[17] Hokenson MJ, Uversky VN, Goers J, Yamin G, Munishkina LA, Fink AL. Role of individual methionines in the fibrillation of methionine-oxidized alpha-synuclein. Biochemistry. 2004;**43**(15):4621-4633. DOI: 10.1021/bi049979h

[18] Leong SL, Pham CL, Galatis D, Fodero-Tavoletti MT, Perez K, Hill AF, et al. Formation of dopamine-mediated alpha-synuclein-soluble oligomers requires methionine oxidation. Free Radical Biology & Medicine. 2009;**46**(10):1328-1337. DOI: 10.1016/j.freeradbiomed.2009.02.009

[19] Elmallah MI, Borgmeyer U, Betzel C, Redecke L. Impact of methionine oxidation as an initial event on the pathway of human prion protein conversion. Prion. 2013;**7**(5):404-411. DOI: 10.4161/pri.26745

[20] Younan ND, Nadal RC, Davies P, Brown DR, Viles JH. Methionine oxidation perturbs the structural core of the prion protein and suggests a generic misfolding pathway. The Journal of Biological Chemistry. 2012;**287**(34):28263-28275. DOI: 10.1074/jbc.M112.354779

[21] Coskun P, Wyrembak J, Schriner SE, Chen HW, Marciniack C, Laferla F, et al. A mitochondrial etiology of Alzheimer and Parkinson disease. Biochimica et Biophysica Acta. 2012;**1820**(5):553-564. DOI: 10.1016/j.bbagen.2011.08.008

[22] Borger E, Aitken L, Muirhead KE, Allen ZE, Ainge JA, Conway SJ, et al. Mitochondrial beta-amyloid in Alzheimer's disease. Biochemical Society Transactions. 2011;**39**(4):868-873. DOI: 10.1042/BST0390868

[23] Siskova Z, Mahad DJ, Pudney C, Campbell G, Cadogan M, Asuni A, et al. Morphological and functional abnormalities in mitochondria associated with synaptic degeneration in prion disease. The American Journal of Pathology. 2010;**177**(3):1411-1421. DOI: 10.2353/ajpath.2010.091037

[24] Brown DR, Qin K, Herms JW, Madlung A, Manson J, Strome R, et al. The cellular prion protein binds copper in vivo. Nature. 1997;**390**(6661):684-687. DOI: 10.1038/37783

[25] Jones CE, Abdelraheim SR, Brown DR, Viles JH. Preferential Cu^{2+} coordination by His96 and His111 induces beta-sheet formation in the unstructured amyloidogenic region of the prion protein. The Journal of Biological Chemistry. 2004;**279**(31):32018-32027. DOI: 10.1074/jbc.M403467200

[26] Zeng L, Zou W, Wang G. Cellular prion protein (PrP(C)) and its role in stress responses. International Journal of Clinical and Experimental Medicine. 2015;**8**(5):8042-8050

[27] Thackray AM, Knight R, Haswell SJ, Bujdoso R, Brown DR. Metal imbalance and compromised antioxidant function are early changes in prion disease. The Biochemical Journal. 2002;**362** (Pt 1):253-258

[28] Brown DR. Neurodegeneration and oxidative stress: Prion disease results from loss of antioxidant defence. Folia Neuropathologica. 2005;**43**(4):229-243

[29] Milhavet O, Lehmann S. Oxidative stress and the prion protein in transmissible spongiform

encephalopathies. Brain Research Brain Research Reviews. 2002;**38**(3):328-339

[30] Nadal RC, Abdelraheim SR, Brazier MW, Rigby SE, Brown DR, Viles JH. Prion protein does not redox-silence Cu^{2+}, but is a sacrificial quencher of hydroxyl radicals. Free Radical Biology & Medicine. 2007;**42**(1):79-89. DOI: 10.1016/j.freeradbiomed.2006.09.019

[31] Alim MA, Hossain MS, Arima K, Takeda K, Izumiyama Y, Nakamura M, et al. Tubulin seeds alpha-synuclein fibril formation. The Journal of Biological Chemistry. 2002;**277**(3): 2112-2117. DOI: 10.1074/jbc.M102981200

[32] Esteves AR, Arduino DM, Swerdlow RH, Oliveira CR, Cardoso SM. Microtubule depolymerization potentiates alpha-synuclein oligomerization. Frontiers in Aging Neuroscience. 2010;**1**:5. DOI: 10.3389/neuro.24.005.2009

[33] Kim M, Jung W, Lee IH, Bhak G, Paik SR, Hahn JS. Impairment of microtubule system increases alpha-synuclein aggregation and toxicity. Biochemical and Biophysical Research Communications. 2008;**365**(4):628-635. DOI: 10.1016/j.bbrc.2007.11.020

[34] Lee HJ, Khoshaghideh F, Lee S, Lee SJ. Impairment of microtubule-dependent trafficking by overexpression of alpha-synuclein. The European Journal of Neuroscience. 2006;**24**(11):3153-3162. DOI: 10.1111/j.1460-9568.2006.05210.x

[35] Nakayama K, Suzuki Y, Yazawa I. Microtubule depolymerization suppresses alpha-synuclein accumulation in a mouse model of multiple system atrophy. The American Journal of Pathology. 2009;**174**(4):1471-1480. DOI: 10.2353/ajpath.2009.080503

[36] Nakayama K, Suzuki Y, Yazawa I. Binding of neuronal alpha-synuclein to beta-III tubulin and accumulation in a model of multiple system atrophy. Biochemical and Biophysical Research Communications. 2012;**417**(4):1170-1175. DOI: 10.1016/j.bbrc.2011.12.092

[37] Zhou RM, Huang YX, Li XL, Chen C, Shi Q, Wang GR, et al. Molecular interaction of alpha-synuclein with tubulin influences on the polymerization of microtubule in vitro and structure of microtubule in cells. Molecular Biology Reports. 2010;**37**(7):3183-3192. DOI: 10.1007/s11033-009-9899-2

[38] Eckert A, Nisbet R, Grimm A, Gotz J. March separate, strike together-role of phosphorylated TAU in mitochondrial dysfunction in Alzheimer's disease. Biochimica et Biophysica Acta. 2014;**1842**(8):1258-1266. DOI: 10.1016/j.bbadis.2013.08.013

[39] Prots I, Veber V, Brey S, Campioni S, Buder K, Riek R, et al. Alpha-synuclein oligomers impair neuronal microtubule-kinesin interplay. The Journal of Biological Chemistry. 2013;**288**(30):21742-21754. DOI: 10.1074/jbc.M113.451815

[40] Schapira AH. Mitochondria in the aetiology and pathogenesis of Parkinson's disease. Lancet Neurology. 2008;**7**(1):97-109. DOI: 10.1016/S1474-4422(07)70327-7

[41] Winslow AR, Chen CW, Corrochano S, Acevedo-Arozena A, Gordon DE, Peden AA, et al. Alpha-synuclein impairs macroautophagy: Implications for Parkinson's disease. The Journal of Cell Biology. 2010;**190**(6): 1023-1037. DOI: 10.1083/jcb.201003122

[42] Lynch-Day MA, Mao K, Wang K, Zhao M, Klionsky DJ. The role of autophagy in Parkinson's disease. Cold Spring Harbor Perspectives in Medicine. 2012;**2**(4):a009357. DOI: 10.1101/cshperspect.a009357

[43] Casley CS, Canevari L, Land JM, Clark JB, Sharpe MA. Beta-amyloid

inhibits integrated mitochondrial respiration and key enzyme activities. Journal of Neurochemistry. 2002;**80**(1):91-100

[44] Yang DS, Stavrides P, Mohan PS, Kaushik S, Kumar A, Ohno M, et al. Reversal of autophagy dysfunction in the TgCRND8 mouse model of Alzheimer's disease ameliorates amyloid pathologies and memory deficits. Brain. 2011;**134**(Pt 1):258-277. DOI: 10.1093/brain/awq341

[45] Dagda RK, Zhu J, Kulich SM, Chu CT. Mitochondrially localized ERK2 regulates mitophagy and autophagic cell stress: Implications for Parkinson's disease. Autophagy. 2008;**4**(6):770-782

[46] Kubli DA, Gustafsson AB. Mitochondria and mitophagy: The yin and yang of cell death control. Circulation Research. 2012;**111**(9):1208-1221. DOI: 10.1161/CIRCRESAHA.112.265819

[47] Yan J, Feng Z, Liu J, Shen W, Wang Y, Wertz K, et al. Enhanced autophagy plays a cardinal role in mitochondrial dysfunction in type 2 diabetic Goto-Kakizaki (GK) rats: Ameliorating effects of (−)-epigallocatechin-3-gallate. The Journal of Nutritional Biochemistry. 2012;**23**(7):716-724. DOI: 10.1016/j.jnutbio.2011.03.014

[48] Zhu JH, Horbinski C, Guo F, Watkins S, Uchiyama Y, Chu CT. Regulation of autophagy by extracellular signal-regulated protein kinases during 1-methyl-4-phenylpyridinium-induced cell death. The American Journal of Pathology. 2007;**170**(1):75-86. DOI: 10.2353/ajpath.2007.060524

[49] Rhein V, Song X, Wiesner A, Ittner LM, Baysang G, Meier F, et al. Amyloid-beta and tau synergistically impair the oxidative phosphorylation system in triple transgenic Alzheimer's disease mice. Proceedings of the National Academy of Sciences of the United States of America. 2009;**106**(47): 20057-20062. DOI: 10.1073/pnas.0905529106

[50] Liberski PP, Sikorska B, Bratosiewicz-Wasik J, Gajdusek DC, Brown P. Neuronal cell death in transmissible spongiform encephalopathies (prion diseases) revisited: From apoptosis to autophagy. The International Journal of Biochemistry & Cell Biology. 2004;**36**(12):2473-2490. DOI: 10.1016/j.biocel.2004.04.016

[51] Liberski PP, Brown DR, Sikorska B, Caughey B, Brown P. Cell death and autophagy in prion diseases (transmissible spongiform encephalopathies). Folia Neuropathologica. 2008;**46**(1):1-25

[52] Liberski PP, Gajdusek DC, Brown P. How do neurons degenerate in prion diseases or transmissible spongiform encephalopathies (TSEs): Neuronal autophagy revisited. Acta Neurobiologiae Experimentalis. 2002;**62**(3):141-147

[53] Shi Y. Caspase activation: Revisiting the induced proximity model. Cell. 2004;**117**(7):855-858. DOI: 10.1016/j.cell.2004.06.007

[54] Shi Y. Mechanisms of caspase activation and inhibition during apoptosis. Molecular Cell. 2002;**9**(3):459-470

[55] Sun X, Wu Y, Chen B, Zhang Z, Zhou W, Tong Y, et al. Regulator of calcineurin 1 (RCAN1) facilitates neuronal apoptosis through caspase-3 activation. The Journal of Biological Chemistry. 2011;**286**(11):9049-9062. DOI: 10.1074/jbc.M110.177519

[56] Eckert A, Marques CA, Keil U, Schussel K, Muller WE. Increased apoptotic cell death in sporadic and genetic Alzheimer's disease. Annals of the New York Academy of Sciences. 2003;**1010**:604-609

[57] Santos LE, Ferreira ST. Crosstalk between endoplasmic reticulum stress

and brain inflammation in Alzheimer's disease. Neuropharmacology. 2018;**136**(Pt B):350-360. DOI: 10.1016/j.neuropharm.2017.11.016

[58] Levine B, Mizushima N, Virgin HW. Autophagy in immunity and inflammation. Nature. 2011;**469**(7330):323-335. DOI: 10.1038/nature09782

[59] Salminen A, Kaarniranta K, Kauppinen A. Inflammaging: Disturbed interplay between autophagy and inflammasomes. Aging. 2012;**4**(3):166-175. DOI: 10.18632/aging.100444

[60] Saitoh T, Fujita N, Jang MH, Uematsu S, Yang BG, Satoh T, et al. Loss of the autophagy protein Atg16L1 enhances endotoxin-induced IL-1beta production. Nature. 2008;**456**(7219):264-268. DOI: 10.1038/nature07383

[61] Nakahira K, Haspel JA, Rathinam VA, Lee SJ, Dolinay T, Lam HC, et al. Autophagy proteins regulate innate immune responses by inhibiting the release of mitochondrial DNA mediated by the NALP3 inflammasome. Nature Immunology. 2011;**12**(3):222-230. DOI: 10.1038/ni.1980

[62] Ghavami S, Shojaei S, Yeganeh B, Ande SR, Jangamreddy JR, Mehrpour M, et al. Autophagy and apoptosis dysfunction in neurodegenerative disorders. Progress in Neurobiology. 2014;**112**:24-49. DOI: 10.1016/j.pneurobio.2013.10.004

[63] Al-Hilaly YK, Williams TL, Stewart-Parker M, Ford L, Skaria E, Cole M, et al. A central role for dityrosine crosslinking of amyloid-beta in Alzheimer's disease. Acta Neuropathologica Communications. 2013;**1**:83. DOI: 10.1186/2051-5960-1-83

[64] Belay ED. Transmissible spongiform encephalopathies in humans. Annual Review of Microbiology. 1999;**53**:283-314. DOI: 10.1146/annurev.micro.53.1.283

[65] Los M, Mozoluk M, Ferrari D, Stepczynska A, Stroh C, Renz A, et al. Activation and caspase-mediated inhibition of PARP: A molecular switch between fibroblast necrosis and apoptosis in death receptor signaling. Molecular Biology of the Cell. 2002;**13**(3):978-988. DOI: 10.1091/mbc.01-05-0272

[66] Eguchi Y, Shimizu S, Tsujimoto Y. Intracellular ATP levels determine cell death fate by apoptosis or necrosis. Cancer Research. 1997;**57**(10):1835-1840

[67] Skulachev VP. Bioenergetic aspects of apoptosis, necrosis and mitoptosis. Apoptosis. 2006;**11**(4):473-485. DOI: 10.1007/s10495-006-5881-9

[68] Milisav I, Suput D, Ribaric S. Unfolded protein response and macroautophagy in Alzheimer's, Parkinson's and prion diseases. Molecules. 2015;**20**(12):22718-22756. DOI: 10.3390/molecules201219865

[69] Jucker M, Walker LC. Self-propagation of pathogenic protein aggregates in neurodegenerative diseases. Nature. 2013;**501**(7465):45-51. DOI: 10.1038/nature12481

[70] Guo JL, Lee VM. Cell-to-cell transmission of pathogenic proteins in neurodegenerative diseases. Nature Medicine. 2014;**20**(2):130-138. DOI: 10.1038/nm.3457

[71] Leist M, Jaattela M. Four deaths and a funeral: From caspases to alternative mechanisms. Nature Reviews Molecular Cell Biology. 2001;**2**(8):589-598. DOI: 10.1038/35085008

[72] Sims JL, Berger SJ, Berger NA. Poly(ADP-ribose) polymerase inhibitors preserve nicotinamide adenine dinucleotide and adenosine 5''-triphosphate pools in DNA-damaged cells: Mechanism of stimulation of unscheduled DNA synthesis. Biochemistry. 1983;**22**(22):5188-5194

[73] Szabo C, Dawson VL. Role of poly(ADP-ribose) synthetase in inflammation and ischaemia-reperfusion. Trends in Pharmacological Sciences. 1998;**19**(7):287-298

[74] Mizushima N. Autophagy: Process and function. Genes & Development. 2007;**21**(22):2861-2873. DOI: 10.1101/gad.1599207

[75] Keil E, Hocker R, Schuster M, Essmann F, Ueffing N, Hoffman B, et al. Phosphorylation of Atg5 by the Gadd45beta-MEKK4-p38 pathway inhibits autophagy. Cell Death and Differentiation. 2013;**20**(2):321-332. DOI: 10.1038/cdd.2012.129

[76] Taneike M, Yamaguchi O, Nakai A, Hikoso S, Takeda T, Mizote I, et al. Inhibition of autophagy in the heart induces age-related cardiomyopathy. Autophagy. 2010;**6**(5):600-606. DOI: 10.4161/auto.6.5.11947

[77] Lepine S, Allegood JC, Edmonds Y, Milstien S, Spiegel S. Autophagy induced by deficiency of sphingosine-1-phosphate phosphohydrolase 1 is switched to apoptosis by calpain-mediated autophagy-related gene 5 (Atg5) cleavage. The Journal of Biological Chemistry. 2011;**286**(52):44380-44390. DOI: 10.1074/jbc.M111.257519

[78] Elfawy HA, Das B. Crosstalk between mitochondrial dysfunction, oxidative stress, and age related neurodegenerative disease: Etiologies and therapeutic strategies. Life Sciences. 2019;**218**:165-184. DOI: 10.1016/j.lfs.2018.12.029

[79] Hamilton A, Zamponi GW, Ferguson SS. Glutamate receptors function as scaffolds for the regulation of beta-amyloid and cellular prion protein signaling complexes. Molecular Brain. 2015;**8**:18. DOI: 10.1186/s13041-015-0107-0

[80] Stys PK, You H, Zamponi GW. Copper-dependent regulation of NMDA receptors by cellular prion protein: Implications for neurodegenerative disorders. The Journal of Physiology. 2012;**590**(6):1357-1368. DOI: 10.1113/jphysiol.2011.225276

[81] Khosravani H, Zhang Y, Tsutsui S, Hameed S, Altier C, Hamid J, et al. Prion protein attenuates excitotoxicity by inhibiting NMDA receptors. The Journal of Cell Biology. 2008;**181**(3):551-565. DOI: 10.1083/jcb.200711002

[82] Shi M, Zhang T, Sun L, Luo Y, Liu DH, Xie ST, et al. Calpain, Atg5 and Bak play important roles in the crosstalk between apoptosis and autophagy induced by influx of extracellular calcium. Apoptosis. 2013;**18**(4):435-451. DOI: 10.1007/s10495-012-0786-2

[83] Xia HG, Zhang L, Chen G, Zhang T, Liu J, Jin M, et al. Control of basal autophagy by calpain1 mediated cleavage of ATG5. Autophagy. 2010;**6**(1):61-66

[84] Zhu X, Messer JS, Wang Y, Lin F, Cham CM, Chang J, et al. Cytosolic HMGB1 controls the cellular autophagy/apoptosis checkpoint during inflammation. The Journal of Clinical Investigation. 2015;**125**(3):1098-1110. DOI: 10.1172/JCI76344

[85] Wei Y, Zou Z, Becker N, Anderson M, Sumpter R, Xiao G, et al. EGFR-mediated Beclin 1 phosphorylation in autophagy suppression, tumor progression, and tumor chemoresistance. Cell. 2013;**154**(6):1269-1284. DOI: 10.1016/j.cell.2013.08.015

[86] Van Humbeeck C, Cornelissen T, Vandenberghe W. Ambra1: A parkin-binding protein involved in mitophagy. Autophagy. 2011;**7**(12):1555-1556

[87] Luo S, Rubinsztein DC. BCL2L11/BIM: A novel molecular link between autophagy and apoptosis. Autophagy. 2013;**9**(1):104-105. DOI: 10.4161/auto.22399

[88] He MX, He YW. CFLAR/c-FLIPL: A star in the autophagy, apoptosis and necroptosis alliance. Autophagy. 2013;**9**(5):791-793. DOI: 10.4161/auto.23785

[89] Hou W, Han J, Lu C, Goldstein LA, Rabinowich H. Autophagic degradation of active caspase-8: A crosstalk mechanism between autophagy and apoptosis. Autophagy. 2010;**6**(7):891-900. DOI: 10.4161/auto.6.7.13038

[90] Cho DH, Jo YK, Hwang JJ, Lee YM, Roh SA, Kim JC. Caspase-mediated cleavage of ATG6/Beclin-1 links apoptosis to autophagy in HeLa cells. Cancer Letters. 2009;**274**(1):95-100. DOI: 10.1016/j.canlet.2008.09.004

[91] Wirawan E, Vande Walle L, Kersse K, Cornelis S, Claerhout S, Vanoverberghe I, et al. Caspase-mediated cleavage of Beclin-1 inactivates Beclin-1-induced autophagy and enhances apoptosis by promoting the release of proapoptotic factors from mitochondria. Cell Death & Disease. 2010;**1**:e18. DOI: 10.1038/cddis.2009.16

[92] Luo S, Rubinsztein DC. Apoptosis blocks Beclin 1-dependent autophagosome synthesis: An effect rescued by Bcl-xL. Cell Death and Differentiation. 2010;**17**(2):268-277. DOI: 10.1038/cdd.2009.121

[93] Kang R, Zeh HJ, Lotze MT, Tang D. The Beclin 1 network regulates autophagy and apoptosis. Cell Death and Differentiation. 2011;**18**(4):571-580. DOI: 10.1038/cdd.2010.191

[94] Zhou F, Yang Y, Xing D. Bcl-2 and Bcl-xL play important roles in the crosstalk between autophagy and apoptosis. The FEBS Journal. 2011;**278**(3):403-413. DOI: 10.1111/j.1742-4658.2010.07965.x

[95] Youle RJ, Strasser A. The BCL-2 protein family: Opposing activities that mediate cell death. Nature Reviews Molecular Cell Biology. 2008;**9**(1):47-59. DOI: 10.1038/nrm2308

[96] Wei MC, Zong WX, Cheng EH, Lindsten T, Panoutsakopoulou V, Ross AJ, et al. Proapoptotic BAX and BAK: A requisite gateway to mitochondrial dysfunction and death. Science. 2001;**292**(5517):727-730. DOI: 10.1126/science.1059108

[97] Zong WX, Lindsten T, Ross AJ, MacGregor GR, Thompson CB. BH3-only proteins that bind pro-survival Bcl-2 family members fail to induce apoptosis in the absence of Bax and Bak. Genes & Development. 2001;**15**(12):1481-1486. DOI: 10.1101/gad.897601

[98] Kang MH, Reynolds CP. Bcl-2 inhibitors: Targeting mitochondrial apoptotic pathways in cancer therapy. Clinical Cancer Research. 2009;**15**(4):1126-1132. DOI: 10.1158/1078-0432.CCR-08-0144

[99] Danial NN. BCL-2 family proteins: Critical checkpoints of apoptotic cell death. Clinical Cancer Research. 2007;**13**(24):7254-7263. DOI: 10.1158/1078-0432.Ccr-07-1598

[100] Cheng EH, Wei MC, Weiler S, Flavell RA, Mak TW, Lindsten T, et al. BCL-2, BCL-X(L) sequester BH3 domain-only molecules preventing BAX- and BAK-mediated mitochondrial apoptosis. Molecular Cell. 2001;**8**(3):705-711

[101] Willis SN, Chen L, Dewson G, Wei A, Naik E, Fletcher JI, et al. Proapoptotic Bak is sequestered by Mcl-1 and Bcl-xL, but not Bcl-2, until displaced by BH3-only proteins. Genes & Development. 2005;**19**(11):1294-1305. DOI: 10.1101/gad.1304105

[102] Werner AB, de Vries E, Tait SW, Bontjer I, Borst J. Bcl-2 family member Bfl-1/A1 sequesters truncated bid to inhibit is collaboration with pro-apoptotic Bak or Bax. The

Journal of Biological Chemistry. 2002;**277**(25):22781-22788. DOI: 10.1074/jbc.M201469200

[103] Kuwana T, Bouchier-Hayes L, Chipuk JE, Bonzon C, Sullivan BA, Green DR, et al. BH3 domains of BH3-only proteins differentially regulate Bax-mediated mitochondrial membrane permeabilization both directly and indirectly. Molecular Cell. 2005;**17**(4): 525-535. DOI: 10.1016/j.molcel.2005.02.003

[104] Kim H, Rafiuddin-Shah M, Tu HC, Jeffers JR, Zambetti GP, Hsieh JJ, et al. Hierarchical regulation of mitochondrion-dependent apoptosis by BCL-2 subfamilies. Nature Cell Biology. 2006;**8**(12):1348-1358. DOI: 10.1038/ncb1499

[105] Erlich S, Mizrachy L, Segev O, Lindenboim L, Zmira O, Adi-Harel S, et al. Differential interactions between Beclin 1 and Bcl-2 family members. Autophagy. 2007;**3**(6):561-568

[106] Maiuri MC, Criollo A, Tasdemir E, Vicencio JM, Tajeddine N, Hickman JA, et al. BH3-only proteins and BH3 mimetics induce autophagy by competitively disrupting the interaction between Beclin 1 and Bcl-2/Bcl-X(L). Autophagy. 2007;**3**(4):374-376

[107] Zalckvar E, Berissi H, Eisenstein M, Kimchi A. Phosphorylation of Beclin 1 by DAP-kinase promotes autophagy by weakening its interactions with Bcl-2 and Bcl-XL. Autophagy. 2009;**5**(5):720-722

[108] Wei Y, Sinha S, Levine B. Dual role of JNK1-mediated phosphorylation of Bcl-2 in autophagy and apoptosis regulation. Autophagy. 2008;**4**(7):949-951

[109] Kitamura Y, Shimohama S, Kamoshima W, Ota T, Matsuoka Y, Nomura Y, et al. Alteration of proteins regulating apoptosis, Bcl-2, Bcl-x, Bax, Bak, Bad, ICH-1 and CPP32, in Alzheimer's disease. Brain Research. 1998;**780**(2):260-269

[110] Roucou X, Gains M, LeBlanc AC. Neuroprotective functions of prion protein. Journal of Neuroscience Research. 2004;**75**(2):153-161. DOI: 10.1002/jnr.10864

[111] Roucou X, Giannopoulos PN, Zhang Y, Jodoin J, Goodyer CG, LeBlanc A. Cellular prion protein inhibits proapoptotic Bax conformational change in human neurons and in breast carcinoma MCF-7 cells. Cell Death and Differentiation. 2005;**12**(7):783-795. DOI: 10.1038/sj.cdd.4401629

[112] Roucou X, Guo Q, Zhang Y, Goodyer CG, LeBlanc AC. Cytosolic prion protein is not toxic and protects against Bax-mediated cell death in human primary neurons. The Journal of Biological Chemistry. 2003;**278**(42):40877-40881. DOI: 10.1074/jbc.M306177200

[113] Rehker J, Rodhe J, Nesbitt RR, Boyle EA, Martin BK, Lord J, et al. Caspase-8, association with Alzheimer's disease and functional analysis of rare variants. PLoS One. 2017;**12**(10):e0185777. DOI: 10.1371/journal.pone.0185777

[114] Feng S, Yang Y, Mei Y, Ma L, Zhu DE, Hoti N, et al. Cleavage of RIP3 inactivates its caspase-independent apoptosis pathway by removal of kinase domain. Cellular Signalling. 2007;**19**(10):2056-2067. DOI: 10.1016/j.cellsig.2007.05.016

[115] O'Donnell MA, Perez-Jimenez E, Oberst A, Ng A, Massoumi R, Xavier R, et al. Caspase 8 inhibits programmed necrosis by processing CYLD. Nature Cell Biology. 2011;**13**(12):1437-1442. DOI: 10.1038/ncb2362

[116] Rebe C, Cathelin S, Launay S, Filomenko R, Prevotat L, L'Ollivier C, et al. Caspase-8 prevents sustained

activation of NF-kappaB in monocytes undergoing macrophagic differentiation. Blood. 2007;**109**(4):1442-1450. DOI: 10.1182/blood-2006-03-011585

[117] Moquin DM, McQuade T, Chan FK. CYLD deubiquitinates RIP1 in the TNFalpha-induced necrosome to facilitate kinase activation and programmed necrosis. PLoS One. 2013;**8**(10):e76841. DOI: 10.1371/journal.pone.0076841

[118] Salvesen GS, Walsh CM. Functions of caspase 8: The identified and the mysterious. Seminars in Immunology. 2014;**26**(3):246-252. DOI: 10.1016/j.smim.2014.03.005

[119] Bjorkblom B, Vainio JC, Hongisto V, Herdegen T, Courtney MJ, Coffey ET. All JNKs can kill, but nuclear localization is critical for neuronal death. The Journal of Biological Chemistry. 2008;**283**(28):19704-19713. DOI: 10.1074/jbc.M707744200

[120] Dhanasekaran DN, Reddy EP. JNK-signaling: A multiplexing hub in programmed cell death. Genes & Cancer. 2017;**8**(9-10):682-694. DOI: 10.18632/genesandcancer.155

[121] Davis RJ. Signal transduction by the JNK group of MAP kinases. Cell. 2000;**103**(2):239-252

[122] Chang L, Karin M. Mammalian MAP kinase signalling cascades. Nature. 2001;**410**(6824):37-40. DOI: 10.1038/35065000

[123] Dhanasekaran DN, Reddy EP. JNK signaling in apoptosis. Oncogene. 2008;**27**(48):6245-6251. DOI: 10.1038/onc.2008.301

[124] Turjanski AG, Vaque JP, Gutkind JS. MAP kinases and the control of nuclear events. Oncogene. 2007;**26**(22):3240-3253. DOI: 10.1038/sj.onc.1210415

[125] Fan M, Chambers TC. Role of mitogen-activated protein kinases in the response of tumor cells to chemotherapy. Drug Resistance Updates. 2001;**4**(4):253-267. DOI: 10.1054/drup.2001.0214

[126] Johnson GL, Nakamura K. The c-Jun kinase/stress-activated pathway: Regulation, function and role in human disease. Biochimica et Biophysica Acta. 2007;**1773**(8):1341-1348. DOI: 10.1016/j.bbamcr.2006.12.009

[127] Fuchs SY, Adler V, Buschmann T, Yin Z, Wu X, Jones SN, et al. JNK targets p53 ubiquitination and degradation in nonstressed cells. Genes & Development. 1998;**12**(17):2658-2663

[128] Oleinik NV, Krupenko NI, Krupenko SA. Cooperation between JNK1 and JNK2 in activation of p53 apoptotic pathway. Oncogene. 2007;**26**(51):7222-7230. DOI: 10.1038/sj.onc.1210526

[129] Tournier C, Hess P, Yang DD, Xu J, Turner TK, Nimnual A, et al. Requirement of JNK for stress-induced activation of the cytochrome c-mediated death pathway. Science. 2000;**288**(5467):870-874

[130] Lei K, Nimnual A, Zong WX, Kennedy NJ, Flavell RA, Thompson CB, et al. The Bax subfamily of Bcl2-related proteins is essential for apoptotic signal transduction by c-Jun NH(2)-terminal kinase. Molecular and Cellular Biology. 2002;**22**(13):4929-4942

[131] Mandal M, Olson DJ, Sharma T, Vadlamudi RK, Kumar R. Butyric acid induces apoptosis by up-regulating Bax expression via stimulation of the c-Jun N-terminal kinase/activation protein-1 pathway in human colon cancer cells. Gastroenterology. 2001;**120**(1):71-78

[132] Papadakis ES, Finegan KG, Wang X, Robinson AC, Guo C, Kayahara M, et al. The regulation of Bax by c-Jun N-terminal protein kinase (JNK) is a prerequisite

to the mitochondrial-induced apoptotic pathway. FEBS Letters. 2006;**580**(5):1320-1326. DOI: 10.1016/j.febslet.2006.01.053

[133] Tsuruta F, Sunayama J, Mori Y, Hattori S, Shimizu S, Tsujimoto Y, et al. JNK promotes Bax translocation to mitochondria through phosphorylation of 14-3-3 proteins. The EMBO Journal. 2004;**23**(8):1889-1899. DOI: 10.1038/sj.emboj.7600194

[134] Donovan N, Becker EB, Konishi Y, Bonni A. JNK phosphorylation and activation of BAD couples the stress-activated signaling pathway to the cell death machinery. The Journal of Biological Chemistry. 2002;**277**(43):40944-40949. DOI: 10.1074/jbc.M206113200

[135] Wang XT, Pei DS, Xu J, Guan QH, Sun YF, Liu XM, et al. Opposing effects of Bad phosphorylation at two distinct sites by Akt1 and JNK1/2 on ischemic brain injury. Cellular Signalling. 2007;**19**(9):1844-1856. DOI: 10.1016/j.cellsig.2007.04.005

[136] Adams JM, Cory S. Bcl-2-regulated apoptosis: Mechanism and therapeutic potential. Current Opinion in Immunology. 2007;**19**(5):488-496. DOI: 10.1016/j.coi.2007.05.004

[137] Lei K, Davis RJ. JNK phosphorylation of Bim-related members of the Bcl2 family induces Bax-dependent apoptosis. Proceedings of the National Academy of Sciences of the United States of America. 2003;**100**(5):2432-2437. DOI: 10.1073/pnas.0438011100

[138] Putcha GV, Le S, Frank S, Besirli CG, Clark K, Chu B, et al. JNK-mediated BIM phosphorylation potentiates BAX-dependent apoptosis. Neuron. 2003;**38**(6):899-914

[139] Marani M, Tenev T, Hancock D, Downward J, Lemoine NR. Identification of novel isoforms of the BH3 domain protein Bim which directly activate Bax to trigger apoptosis. Molecular and Cellular Biology. 2002;**22**(11):3577-3589

[140] Letai A, Bassik MC, Walensky LD, Sorcinelli MD, Weiler S, Korsmeyer SJ. Distinct BH3 domains either sensitize or activate mitochondrial apoptosis, serving as prototype cancer therapeutics. Cancer Cell. 2002;**2**(3):183-192

[141] Willis SN, Fletcher JI, Kaufmann T, van Delft MF, Chen L, Czabotar PE, et al. Apoptosis initiated when BH3 ligands engage multiple Bcl-2 homologs, not Bax or Bak. Science. 2007;**315**(5813):856-859. DOI: 10.1126/science.1133289

[142] Hubner A, Barrett T, Flavell RA, Davis RJ. Multisite phosphorylation regulates Bim stability and apoptotic activity. Molecular Cell. 2008;**30**(4):415-425. DOI: 10.1016/j.molcel.2008.03.025

[143] Yamamoto K, Ichijo H, Korsmeyer SJ. BCL-2 is phosphorylated and inactivated by an ASK1/Jun N-terminal protein kinase pathway normally activated at G(2)/M. Molecular and Cellular Biology. 1999;**19**(12):8469-8478

[144] Walczak H. Death receptor-ligand systems in cancer, cell death, and inflammation. Cold Spring Harbor Perspectives in Biology. 2013;**5**(5):a008698. DOI: 10.1101/cshperspect.a008698

[145] Papa S, Zazzeroni F, Pham CG, Bubici C, Franzoso G. Linking JNK signaling to NF-kappaB: A key to survival. Journal of Cell Science. 2004;**117**(Pt 22):5197-5208. DOI: 10.1242/jcs.01483

[146] Higuchi H, Grambihler A, Canbay A, Bronk SF, Gores GJ. Bile acids up-regulate death receptor 5/TRAIL-receptor 2 expression via a c-Jun

N-terminal kinase-dependent pathway involving Sp1. The Journal of Biological Chemistry. 2004;**279**(1):51-60. DOI: 10.1074/jbc.M309476200

[147] Zou W, Liu X, Yue P, Zhou Z, Sporn MB, Lotan R, et al. c-Jun NH2-terminal kinase-mediated up-regulation of death receptor 5 contributes to induction of apoptosis by the novel synthetic triterpenoid methyl-2-cyano-3,12-dioxooleana-1, 9-dien-28-oate in human lung cancer cells. Cancer Research. 2004;**64**(20):7570-7578. DOI: 10.1158/0008-5472.CAN-04-1238

[148] Jin HO, Park IC, An S, Lee HC, Woo SH, Hong YJ, et al. Up-regulation of Bak and Bim via JNK downstream pathway in the response to nitric oxide in human glioblastoma cells. Journal of Cellular Physiology. 2006;**206**(2): 477-486. DOI: 10.1002/jcp.20488

[149] Chen YJ, Liu WH, Kao PH, Wang JJ, Chang LS. Involvement of p38 MAPK- and JNK-modulated expression of Bcl-2 and Bax in Naja nigricollis CMS-9-induced apoptosis of human leukemia K562 cells. Toxicon. 2010;**55**(7):1306-1316. DOI: 10.1016/j.toxicon.2010.01.024

[150] Deng Y, Ren X, Yang L, Lin Y, Wu X. A JNK-dependent pathway is required for TNFalpha-induced apoptosis. Cell. 2003;**115**(1):61-70

[151] Silke J, Meier P. Inhibitor of apoptosis (IAP) proteins-modulators of cell death and inflammation. Cold Spring Harbor Perspectives in Biology. 2013;**5**(2):1-19. DOI: 10.1101/cshperspect.a008730

[152] Ploia C, Antoniou X, Sclip A, Grande V, Cardinetti D, et al. JNK plays a key role in tau hyperphosphorylation in Alzheimer's disease models. Journal of Alzheimer's Disease. 2011;**26**(2):315-329. DOI: 10.3233/JAD-2011-110320. PubMed PMID: 21628793

[153] Morishima Y, Gotoh Y, Zieg J, Barrett T, Takano H, Flavell R, et al. Beta-amyloid induces neuronal apoptosis via a mechanism that involves the c-Jun N-terminal kinase pathway and the induction of Fas ligand. The Journal of Neuroscience. 2001;**21**(19):7551-7560

[154] Minogue AM, Schmid AW, Fogarty MP, Moore AC, Campbell VA, Herron CE, et al. Activation of the c-Jun N-terminal kinase signaling cascade mediates the effect of amyloid-beta on long term potentiation and cell death in hippocampus: A role for interleukin-1beta? The Journal of Biological Chemistry. 2003;**278**(30):27971-27980. DOI: 10.1074/jbc.M302530200

[155] Zhu X, Raina AK, Rottkamp CA, Aliev G, Perry G, Boux H, et al. Activation and redistribution of c-Jun N-terminal kinase/stress activated protein kinase in degenerating neurons in Alzheimer's disease. Journal of Neurochemistry. 2001;**76**(2):435-441

[156] Ferrer I, Blanco R, Carmona M, Puig B. Phosphorylated mitogen-activated protein kinase (MAPK/ERK-P), protein kinase of 38 kDa (p38-P), stress-activated protein kinase (SAPK/JNK-P), and calcium/calmodulin-dependent kinase II (CaM kinase II) are differentially expressed in tau deposits in neurons and glial cells in tauopathies. Journal of Neural Transmission (Vienna). 2001;**108**(12):1397-1415. DOI: 10.1007/s007020100016

[157] Bellucci A, Rosi MC, Grossi C, Fiorentini A, Luccarini I, Casamenti F. Abnormal processing of tau in the brain of aged TgCRND8 mice. Neurobiology of Disease. 2007;**27**(3):328-338. DOI: 10.1016/j.nbd.2007.06.008

[158] Ma QL, Yang F, Rosario ER, Ubeda OJ, Beech W, Gant DJ, et al. Beta-amyloid oligomers induce phosphorylation of tau and inactivation of insulin receptor substrate via

c-Jun N-terminal kinase signaling: Suppression by omega-3 fatty acids and curcumin. The Journal of Neuroscience. 2009;**29**(28):9078-9089. DOI: 10.1523/JNEUROSCI.1071-09.2009

[159] Vogel J, Anand VS, Ludwig B, Nawoschik S, Dunlop J, Braithwaite SP. The JNK pathway amplifies and drives subcellular changes in tau phosphorylation. Neuropharmacology. 2009;**57**(5-6):539-550. DOI: 10.1016/j.neuropharm.2009.07.021

[160] Drewes G, Ebneth A, Mandelkow EM. MAPs, MARKs and microtubule dynamics. Trends in Biochemical Sciences. 1998;**23**(8):307-311

[161] Kosik KS, McConlogue L. Microtubule-associated protein function: Lessons from expression in Spodoptera frugiperda cells. Cell Motility and the Cytoskeleton. 1994;**28**(3):195-198. DOI: 10.1002/cm.970280302

[162] Illenberger S, Zheng-Fischhofer Q, Preuss U, Stamer K, Baumann K, Trinczek B, et al. The endogenous and cell cycle-dependent phosphorylation of tau protein in living cells: Implications for Alzheimer's disease. Molecular Biology of the Cell. 1998;**9**(6):1495-1512

[163] Mandelkow EM, Biernat J, Drewes G, Gustke N, Trinczek B, Mandelkow E. Tau domains, phosphorylation, and interactions with microtubules. Neurobiology of Aging. 1995;**16**(3):355-362. Discussion: 362-353

[164] Trojanowski JQ, Lee VM. Phosphorylation of paired helical filament tau in Alzheimer's disease neurofibrillary lesions: Focusing on phosphatases. The FASEB Journal. 1995;**9**(15):1570-1576

[165] Delacourte A, Buee L. Normal and pathological tau proteins as factors for microtubule assembly. International Review of Cytology. 1997;**171**:167-224

[166] Mandelkow EM, Mandelkow E. Tau in Alzheimer's disease. Trends in Cell Biology. 1998;**8**(11):425-427

[167] Linkermann A, Green DR. Necroptosis. The New England Journal of Medicine. 2014;**370**(5):455-465. DOI: 10.1056/NEJMra1310050

[168] Wang Z, Jiang H, Chen S, Du F, Wang X. The mitochondrial phosphatase PGAM5 functions at the convergence point of multiple necrotic death pathways. Cell. 2012;**148**(1-2):228-243. DOI: 10.1016/j.cell.2011.11.030

[169] Cai Z, Jitkaew S, Zhao J, Chiang HC, Choksi S, Liu J, et al. Plasma membrane translocation of trimerized MLKL protein is required for TNF-induced necroptosis. Nature Cell Biology. 2014;**16**(1):55-65. DOI: 10.1038/ncb2883

[170] Sun W, Wu X, Gao H, Yu J, Zhao W, Lu JJ, et al. Cytosolic calcium mediates RIP1/RIP3 complex-dependent necroptosis through JNK activation and mitochondrial ROS production in human colon cancer cells. Free Radical Biology & Medicine. 2017;**108**:433-444. DOI: 10.1016/j.freeradbiomed.2017.04.010

[171] Kamata H, Honda S, Maeda S, Chang L, Hirata H, Karin M. Reactive oxygen species promote TNFalpha-induced death and sustained JNK activation by inhibiting MAP kinase phosphatases. Cell. 2005;**120**(5):649-661. DOI: 10.1016/j.cell.2004.12.041

[172] Lee JS, Li Q, Lee JY, Lee SH, Jeong JH, Lee HR, et al. FLIP-mediated autophagy regulation in cell death control. Nature Cell Biology. 2009;**11**(11):1355-1362. DOI: 10.1038/ncb1980

[173] Fujita K, Motoki K, Tagawa K, Chen X, Hama H, Nakajima K, et al. HMGB1, a pathogenic molecule that induces neurite degeneration

via TLR4-MARCKS, is a potential therapeutic target for Alzheimer's disease. Scientific Reports. 2016;**6**:31895. DOI: 10.1038/srep31895

[174] Paudel YN, Shaikh MF, Chakraborti A, Kumari Y, Aledo-Serrano A, Aleksovska K, et al. HMGB1: A common biomarker and potential target for TBI, neuroinflammation, epilepsy, and cognitive dysfunction. Frontiers in Neuroscience. 2018;**12**:628. DOI: 10.3389/fnins.2018.00628

[175] Heneka MT, Carson MJ, El Khoury J, Landreth GE, Brosseron F, Feinstein DL, et al. Neuroinflammation in Alzheimer's disease. Lancet Neurology. 2015;**14**(4):388-405. DOI: 10.1016/S1474-4422(15)70016-5

[176] Tansey MG, Goldberg MS. Neuroinflammation in Parkinson's disease: Its role in neuronal death and implications for therapeutic intervention. Neurobiology of Disease. 2010;**37**(3):510-518. DOI: 10.1016/j.nbd.2009.11.004

[177] Frank MG, Weber MD, Fonken LK, Hershman SA, Watkins LR, Maier SF. The redox state of the alarmin HMGB1 is a pivotal factor in neuroinflammatory and microglial priming: A role for the NLRP3 inflammasome. Brain, Behavior, and Immunity. 2016;**55**:215-224. DOI: 10.1016/j.bbi.2015.10.009

[178] Takata K, Kitamura Y, Tsuchiya D, Kawasaki T, Taniguchi T, Shimohama S. High mobility group box protein-1 inhibits microglial Abeta clearance and enhances Abeta neurotoxicity. Journal of Neuroscience Research. 2004;**78**(6):880-891. DOI: 10.1002/jnr.20340

[179] Takata K, Takada T, Ito A, Asai M, Tawa M, Saito Y, et al. Microglial amyloid-beta1-40 phagocytosis dysfunction is caused by high-mobility group box protein-1: Implications for the pathological progression of Alzheimer's disease. International Journal of Alzheimer's Disease. 2012;**2012**:685739. DOI: 10.1155/2012/685739

[180] Ma Y, Shi Q, Wang J, Xiao K, Sun J, Lv Y, et al. Reduction of NF-kappaB (p65) in scrapie-infected cultured cells and in the brains of scrapie-infected rodents. ACS Chemical Neuroscience. 2017;**8**(11):2535-2548. DOI: 10.1021/acschemneuro.7b00273

[181] Barger SW, Mattson MP. Induction of neuroprotective kappa B-dependent transcription by secreted forms of the Alzheimer's beta-amyloid precursor. Brain Research Molecular Brain Research. 1996;**40**(1):116-126

[182] Guo Q, Robinson N, Mattson MP. Secreted beta-amyloid precursor protein counteracts the proapoptotic action of mutant presenilin-1 by activation of NF-kappaB and stabilization of calcium homeostasis. The Journal of Biological Chemistry. 1998;**273**(20):12341-12351

[183] Mattson MP, Goodman Y, Luo H, Fu W, Furukawa K. Activation of NF-kappaB protects hippocampal neurons against oxidative stress-induced apoptosis: Evidence for induction of manganese superoxide dismutase and suppression of peroxynitrite production and protein tyrosine nitration. Journal of Neuroscience Research. 1997;**49**(6):681-697. DOI: 10.1002/(SICI)1097-4547(19970915)49:6<681:AID-JNR3>3.0.CO;2-3

[184] Barger SW, Horster D, Furukawa K, Goodman Y, Krieglstein J, Mattson MP. Tumor necrosis factors alpha and beta protect neurons against amyloid beta-peptide toxicity: Evidence for involvement of a kappa B-binding factor and attenuation of peroxide and Ca^{2+} accumulation. Proceedings of the National Academy of Sciences of the United States of America. 1995;**92**(20):9328-9332

[185] Hunot S, Brugg B, Ricard D, Michel PP, Muriel MP, Ruberg M, et al. Nuclear translocation of NF-kappaB is increased in dopaminergic neurons

of patients with parkinson disease. Proceedings of the National Academy of Sciences of the United States of America. 1997;**94**(14):7531-7536

[186] Kihara A, Kabeya Y, Ohsumi Y, Yoshimori T. Beclin-phosphatidylinositol 3-kinase complex functions at the trans-Golgi network. EMBO Reports. 2001;**2**(4):330-335. DOI: 10.1093/embo-reports/kve061

[187] Kang R, Livesey KM, Zeh HJ, Loze MT, Tang D. HMGB1: A novel Beclin 1-binding protein active in autophagy. Autophagy. 2010;**6**(8):1209-1211

[188] Morselli E, Galluzzi L, Kepp O, Marino G, Michaud M, Vitale I, et al. Oncosuppressive functions of autophagy. Antioxidants & Redox Signaling. 2011;**14**(11):2251-2269. DOI: 10.1089/ars.2010.3478

[189] Shih RH, Wang CY, Yang CM. NF-kappaB signaling pathways in neurological inflammation: A mini review. Frontiers in Molecular Neuroscience. 2015;**8**:77. DOI: 10.3389/fnmol.2015.00077

[190] Milne JC, Denu JM. The Sirtuin family: Therapeutic targets to treat diseases of aging. Current Opinion in Chemical Biology. 2008;**12**(1):11-17. DOI: 10.1016/j.cbpa.2008.01.019

[191] Lee IH, Cao L, Mostoslavsky R, Lombard DB, Liu J, Bruns NE, et al. A role for the NAD-dependent deacetylase Sirt1 in the regulation of autophagy. Proceedings of the National Academy of Sciences of the United States of America. 2008;**105**(9):3374-3379. DOI: 10.1073/pnas.0712145105

[192] Narayan N, Lee IH, Borenstein R, Sun J, Wong R, Tong G, et al. The NAD-dependent deacetylase SIRT2 is required for programmed necrosis. Nature. 2012;**492**(7428):199-204. DOI: 10.1038/nature11700

[193] Kim D, Nguyen MD, Dobbin MM, Fischer A, Sananbenesi F, Rodgers JT, et al. SIRT1 deacetylase protects against neurodegeneration in models for Alzheimer's disease and amyotrophic lateral sclerosis. The EMBO Journal. 2007;**26**(13):3169-3179. DOI: 10.1038/sj.emboj.7601758

[194] Julien C, Tremblay C, Emond V, Lebbadi M, Salem N Jr, Bennett DA, et al. Sirtuin 1 reduction parallels the accumulation of tau in Alzheimer disease. Journal of Neuropathology and Experimental Neurology. 2009;**68**(1):48-58. DOI: 10.1097/NEN.0b013e3181922348

[195] Chipuk JE, Kuwana T, Bouchier-Hayes L, Droin NM, Newmeyer DD, Schuler M, et al. Direct activation of Bax by p53 mediates mitochondrial membrane permeabilization and apoptosis. Science. 2004;**303**(5660):1010-1014. DOI: 10.1126/science.1092734

[196] Nakano K, Vousden KH. PUMA, a novel proapoptotic gene, is induced by p53. Molecular Cell. 2001;**7**(3):683-694

[197] Chipuk JE, Bouchier-Hayes L, Kuwana T, Newmeyer DD, Green DR. PUMA couples the nuclear and cytoplasmic proapoptotic function of p53. Science. 2005;**309**(5741):1732-1735. DOI: 10.1126/science.1114297

[198] Mah LY, O'Prey J, Baudot AD, Hoekstra A, Ryan KM. DRAM-1 encodes multiple isoforms that regulate autophagy. Autophagy. 2012;**8**(1):18-28. DOI: 10.4161/auto.8.1.18077

[199] Gao W, Shen Z, Shang L, Wang X. Upregulation of human autophagy-initiation kinase ULK1 by tumor suppressor p53 contributes to DNA-damage-induced cell death. Cell Death and Differentiation. 2011;**18**(10):1598-1607. DOI: 10.1038/cdd.2011.33

[200] Feng Z, Zhang H, Levine AJ, Jin S. The coordinate regulation of the p53 and mTOR pathways in cells. Proceedings of the National Academy of Sciences of the United States of America. 2005;**102**(23):8204-8209. DOI: 10.1073/pnas.0502857102

[201] Crighton D, Wilkinson S, O'Prey J, Syed N, Smith P, Harrison PR, et al. DRAM, a p53-induced modulator of autophagy, is critical for apoptosis. Cell. 2006;**126**(1):121-134. DOI: 10.1016/j.cell.2006.05.034

[202] Merlo P, Frost B, Peng S, Yang YJ, Park PJ, Feany M. p53 prevents neurodegeneration by regulating synaptic genes. Proceedings of the National Academy of Sciences of the United States of America. 2014;**111**(50):18055-18060. DOI: 10.1073/pnas.1419083111

[203] Kitamura Y, Shimohama S, Kamoshima W, Matsuoka Y, Nomura Y, et al. Changes of p53 in the brains of patients with Alzheimer's disease. Biochemical and Biophysical Research Communications. 1997;**232**:418-421

[204] Su JH, Deng G, Cotman CW. Bax protein expression is increased in Alzheimer's brain: Correlations with DNA damage, Bcl-2 expression, and brain pathology. Journal of Neuropathology and Experimental Neurology. 1997;**56**(1):86-93

[205] Farmer KM, Carretero-Murillo M, McAllen S, Sarkar P, Kayed R. Tau and P53 in Alzheimer's disease. Alzheimer's & Dementia. 2017;**13**(7):1505. DOI: 10.1016/j.jalz.2017.07.617

[206] Goodall ML, Fitzwalter BE, Zahedi S, Wu M, Rodriguez D, Mulcahy-Levy JM, et al. The autophagy machinery controls cell death switching between apoptosis and necroptosis. Developmental Cell. 2016;**37**(4):337-349. DOI: 10.1016/j.devcel.2016.04.018

[207] Moscat J, Diaz-Meco MT. p62 at the crossroads of autophagy, apoptosis, and cancer. Cell. 2009;**137**(6):1001-1004. DOI: 10.1016/j.cell.2009.05.023

[208] Salminen A, Kaarniranta K, Haapasalo A, Hiltunen M, Soininen H, Alafuzoff I. Emerging role of p62/sequestosome-1 in the pathogenesis of Alzheimer's disease. Progress in Neurobiology. 2012;**96**(1):87-95. DOI: 10.1016/j.pneurobio.2011.11.005

[209] Komatsu M, Kageyama S, Ichimura Y. p62/SQSTM1/A170: Physiology and pathology. Pharmacological Research. 2012;**66**(6):457-462. DOI: 10.1016/j.phrs.2012.07.004

[210] Zhang YB, Gong JL, Xing TY, Zheng SP, Ding W. Autophagy protein p62/SQSTM1 is involved in HAMLET-induced cell death by modulating apotosis in U87MG cells. Cell Death & Disease. 2013;**4**:e550. DOI: 10.1038/cddis.2013.77

[211] Jin Z, Li Y, Pitti R, Lawrence D, Pham VC, Lill JR, et al. Cullin3-based polyubiquitination and p62-dependent aggregation of caspase-8 mediate extrinsic apoptosis signaling. Cell. 2009;**137**(4):721-735. DOI: 10.1016/j.cell.2009.03.015

[212] Khan SH, Zhao D, Shah SZ, Hassan MF, Zhu T, Song Z, et al. Parkin overexpression ameliorates PrP106-126-induced neurotoxicity via enhanced autophagy in N2a cells. Cellular and Molecular Neurobiology. 2017;**37**(4):717-728. DOI: 10.1007/s10571-016-0407-7

[213] Homma T, Ishibashi D, Nakagaki T, Satoh K, Sano K, Atarashi R, et al. Increased expression of p62/SQSTM1 in prion diseases and its association with pathogenic prion protein. Scientific Reports. 2014;**4**:4504. DOI: 10.1038/srep04504

[214] Katsuragi Y, Ichimura Y, Komatsu M. Regulation of the Keap1–Nrf2

pathway by p62/SQSTM1. Current Opinion in Toxicology. 2016;**1**:54-61. DOI: 10.1016/j.cotox.2016.09.005

[215] Caccamo A, Ferreira E, Branca C, Oddo S. p62 improves AD-like pathology by increasing autophagy. Molecular Psychiatry. 2017;**22**(6): 865-873. DOI: 10.1038/mp.2016.139

[216] Carroll JA, Race B, Williams K, Chesebro B. Toll-like receptor 2 confers partial neuroprotection during prion disease. PLoS One. 2018;**13**(12):e0208559. DOI: 10.1371/journal.pone.0208559

[217] Spinner DS, Cho IS, Park SY, Kim JI, Meeker HC, Ye X, et al. Accelerated prion disease pathogenesis in toll-like receptor 4 signaling-mutant mice. Journal of Virology. 2008;**82**(21):10701-10708. DOI: 10.1128/JVI.00522-08

[218] Walter S, Letiembre M, Liu Y, Heine H, Penke B, Hao W, et al. Role of the toll-like receptor 4 in neuroinflammation in Alzheimer's disease. Cellular Physiology and Biochemistry. 2007;**20**(6):947-956. DOI: 10.1159/000110455

[219] Minoretti P, Gazzaruso C, Vito CD, Emanuele E, Bianchi M, Coen E, et al. Effect of the functional toll-like receptor 4 Asp299Gly polymorphism on susceptibility to late-onset Alzheimer's disease. Neuroscience Letters. 2006;**391**(3):147-149. DOI: 10.1016/j.neulet.2005.08.047

[220] Reed-Geaghan EG, Savage JC, Hise AG, Landreth GE. CD14 and toll-like receptors 2 and 4 are required for fibrillar a{beta}-stimulated microglial activation. The Journal of Neuroscience. 2009;**29**(38):11982-11992. DOI: 10.1523/JNEUROSCI.3158-09.2009

[221] Lotz M, Ebert S, Esselmann H, Iliev AI, Prinz M, Wiazewicz N, et al. Amyloid beta peptide 1-40 enhances the action of toll-like receptor-2 and -4 agonists but antagonizes toll-like receptor-9-induced inflammation in primary mouse microglial cell cultures. Journal of Neurochemistry. 2005;**94**(2):289-298. DOI: 10.1111/j.1471-4159.2005.03188.x

[222] Tahara K, Kim HD, Jin JJ, Maxwell JA, Li L, Fukuchi K. Role of toll-like receptor signalling in Abeta uptake and clearance. Brain. 2006;**129**(Pt 11): 3006-3019. DOI: 10.1093/brain/awl249

[223] Tang SC, Lathia JD, Selvaraj PK, Jo DG, Mughal MR, Cheng A, et al. Toll-like receptor-4 mediates neuronal apoptosis induced by amyloid beta-peptide and the membrane lipid peroxidation product 4-hydroxynonenal. Experimental Neurology. 2008;**213**(1):114-121. DOI: 10.1016/j.expneurol.2008.05.014

[224] Calvo-Rodriguez M, de la Fuente C, Garcia-Durillo M, Garcia-Rodriguez C, Villalobos C, Nunez L. Aging and amyloid beta oligomers enhance TLR4 expression, LPS-induced Ca(2+) responses, and neuron cell death in cultured rat hippocampal neurons. Journal of Neuroinflammation. 2017;**14**(1):24. DOI: 10.1186/s12974-017-0802-0

[225] Gasic-Milenkovic J, Dukic-Stefanovic S, Deuther-Conrad W, Gartner U, Munch G. Beta-amyloid peptide potentiates inflammatory responses induced by lipopolysaccharide, interferon-gamma and "advanced glycation 'endproducts' in a murine microglia cell line. The European Journal of Neuroscience. 2003;**17**(4):813-821

[226] Perez-Pardo P, Dodiya HB, Engen PA, Forsyth CB, Huschens AM, Shaikh M, et al. Role of TLR4 in the gut-brain axis in Parkinson's disease: A translational study from men to mice. Gut. 2018;**68**:829-843. DOI: 10.1136/gutjnl-2018-316844

[227] Perez-Pardo P, Dodiya HB, Broersen LM, Douna H, van Wijk N,

Lopes da Silva S, et al. Gut-brain and brain-gut axis in Parkinson's disease models: Effects of a uridine and fish oil diet. Nutritional Neuroscience. 2018;**21**(6):391-402. DOI: 10.1080/1028415X.2017.1294555

[228] Pan-Montojo F, Anichtchik O, Dening Y, Knels L, Pursche S, Jung R, et al. Progression of Parkinson's disease pathology is reproduced by intragastric administration of rotenone in mice. PLoS One. 2010;**5**(1):e8762. DOI: 10.1371/journal.pone.0008762

[229] Drolet RE, Cannon JR, Montero L, Greenamyre JT. Chronic rotenone exposure reproduces Parkinson's disease gastrointestinal neuropathology. Neurobiology of Disease. 2009;**36**(1): 96-102. DOI: 10.1016/j.nbd.2009.06.017

[230] Clairembault T, Leclair-Visonneau L, Neunlist M, Derkinderen P. Enteric glial cells: New players in Parkinson's disease? Movement Disorders. 2015;**30**(4): 494-498. DOI: 10.1002/mds.25979

[231] Stefanova N, Reindl M, Neumann M, Kahle PJ, Poewe W, Wenning GK. Microglial activation mediates neurodegeneration related to oligodendroglial alpha-synucleinopathy: Implications for multiple system atrophy. Movement Disorders. 2007;**22**(15): 2196-2203. DOI: 10.1002/mds.21671

[232] Drouin-Ouellet J, St-Amour I, Saint-Pierre M, Lamontagne-Proulx J, Kriz J, Barker RA, et al. Toll-like receptor expression in the blood and brain of patients and a mouse model of Parkinson's disease. The International Journal of Neuropsychopharmacology. 2014;**18**(6):1-11. DOI: 10.1093/ijnp/pyu103

[233] Stefanova N, Fellner L, Reindl M, Masliah E, Poewe W, Wenning GK. Toll-like receptor 4 promotes alpha-synuclein clearance and survival of nigral dopaminergic neurons. The American Journal of Pathology. 2011;**179**(2):954-963. DOI: 10.1016/j.ajpath.2011.04.013

[234] Rietdijk CD, Van Wezel RJA, Garssen J, Kraneveld AD. Neuronal toll-like receptors and neuro-immunity in Parkinson's disease, Alzheimer's disease and stroke. Neuroimmunology and Neuroinflammation. 2016;**3**(2):27-37. DOI: 10.20517/2347-8659.2015.28

[235] Heiseke A, Aguib Y, Schatzl HM. Autophagy, prion infection and their mutual interactions. Current Issues in Molecular Biology. 2010;**12**(2):87-97

[236] Ravikumar B, Sarkar S, Davies JE, Futter M, Garcia-Arencibia M, Green-Thompson ZW, et al. Regulation of mammalian autophagy in physiology and pathophysiology. Physiological Reviews. 2010;**90**(4):1383-1435. DOI: 10.1152/physrev.00030.2009

[237] Martinez-Vicente M. Autophagy in neurodegenerative diseases: From pathogenic dysfunction to therapeutic modulation. Seminars in Cell & Developmental Biology. 2015;**40**:115-126. DOI: 10.1016/j.semcdb.2015.03.005

[238] Menzies FM, Fleming A, Caricasole A, Bento CF, Andrews SP, Ashkenazi A, et al. Autophagy and neurodegeneration: Pathogenic mechanisms and therapeutic opportunities. Neuron. 2017;**93**(5): 1015-1034. DOI: 10.1016/j.neuron.2017.01.022

[239] Heitz S, Grant NJ, Bailly Y. Doppel induces autophagic stress in prion protein-deficient Purkinje cells. Autophagy. 2009;**5**(3):422-424

[240] Heitz S, Grant NJ, Leschiera R, Haeberle AM, Demais V, Bombarde G, et al. Autophagy and cell death of Purkinje cells overexpressing Doppel in Ngsk Prnp-deficient mice. Brain Pathology. 2010;**20**(1):119-132. DOI: 10.1111/j.1750-3639.2008.00245.x

[241] Ma Y, Shi Q, Xiao K, Wang J, Chen C, Gao LP, et al. Stimulations of the culture medium of activated microglia and TNF-alpha on a

scrapie-infected cell line decrease the cell viability and induce marked necroptosis that also occurs in the brains from the patients of human prion diseases. ACS Chemical Neuroscience. 2018;**10**:1273-1283. DOI: 10.1021/acschemneuro.8b00354

[242] Wu HJ, Pu JL, Krafft PR, Zhang JM, Chen S. The molecular mechanisms between autophagy and apoptosis: Potential role in central nervous system disorders. Cellular and Molecular Neurobiology. 2015;**35**(1):85-99. DOI: 10.1007/s10571-014-0116-z

[243] Forbes JM, Thorburn DR. Mitochondrial dysfunction in diabetic kidney disease. Nature Reviews. Nephrology. 2018;**14**(5):291-312. DOI: 10.1038/nrneph.2018.9

[244] Szeto HH, Birk AV. Serendipity and the discovery of novel compounds that restore mitochondrial plasticity. Clinical Pharmacology and Therapeutics. 2014;**96**(6):672-683. DOI: 10.1038/clpt.2014.174

[245] Moussaoui S, Obinu MC, Daniel N, Reibaud M, Blanchard V, Imperato A. The antioxidant ebselen prevents neurotoxicity and clinical symptoms in a primate model of Parkinson's disease. Experimental Neurology. 2000;**166**(2):235-245. DOI: 10.1006/exnr.2000.7516

[246] Horvath TL, Diano S, Leranth C, Garcia-Segura LM, Cowley MA, Shanabrough M, et al. Coenzyme Q induces nigral mitochondrial uncoupling and prevents dopamine cell loss in a primate model of Parkinson's disease. Endocrinology. 2003;**144**(7):2757-2760. DOI: 10.1210/en.2003-0163

[247] Levites Y, Weinreb O, Maor G, Youdim MB, Mandel S. Green tea polyphenol (−)-epigallocatechin-3-gallate prevents N-methyl-4-phenyl-1,2,3,6-tetrahydropyridine-induced dopaminergic neurodegeneration. Journal of Neurochemistry. 2001;**78**(5):1073-1082

[248] Karunakaran S, Diwakar L, Saeed U, Agarwal V, Ramakrishnan S, Iyengar S, et al. Activation of apoptosis signal regulating kinase 1 (ASK1) and translocation of death-associated protein, Daxx, in substantia nigra pars compacta in a mouse model of Parkinson's disease: Protection by alpha-lipoic acid. The FASEB Journal. 2007;**21**(9):2226-2236. DOI: 10.1096/fj.06-7580com

[249] Kwon HJ, Cha MY, Kim D, Kim DK, Soh M, Shin K, et al. Mitochondria-targeting ceria nanoparticles as antioxidants for Alzheimer's disease. ACS Nano. 2016;**10**(2):2860-2870. DOI: 10.1021/acsnano.5b08045

[250] Sweeney G, Song J. The association between PGC-1alpha and Alzheimer's disease. Anatomy & Cell Biology. 2016;**49**(1):1-6. DOI: 10.5115/acb.2016.49.1.1

[251] Cortes CJ, Qin K, Cook J, Solanki A, Mastrianni JA. Rapamycin delays disease onset and prevents PrP plaque deposition in a mouse model of Gerstmann-Straussler-Scheinker disease. The Journal of Neuroscience. 2012;**32**(36):12396-12405. DOI: 10.1523/JNEUROSCI.6189-11.2012

[252] Jiang T, Yu JT, Zhu XC, Tan MS, Wang HF, Cao L, et al. Temsirolimus promotes autophagic clearance of amyloid-beta and provides protective effects in cellular and animal models of Alzheimer's disease. Pharmacological Research. 2014;**81**:54-63. DOI: 10.1016/j.phrs.2014.02.008

[253] Menzies FM, Huebener J, Renna M, Bonin M, Riess O, Rubinsztein DC. Autophagy induction reduces mutant ataxin-3 levels and toxicity in a mouse model of spinocerebellar ataxia type 3. Brain. 2010;**133**(Pt 1):93-104. DOI: 10.1093/brain/awp292

[254] Ozcelik S, Fraser G, Castets P, Schaeffer V, Skachokova Z, Breu K, et al. Rapamycin attenuates the progression of tau pathology in P301S tau transgenic mice. PLoS One. 2013;**8**(5):e62459. DOI: 10.1371/journal.pone.0062459

[255] Ravikumar B, Vacher C, Berger Z, Davies JE, Luo S, Oroz LG, et al. Inhibition of mTOR induces autophagy and reduces toxicity of polyglutamine expansions in fly and mouse models of Huntington disease. Nature Genetics. 2004;**36**(6):585-595. DOI: 10.1038/ng1362

[256] Sarkar S, Krishna G, Imarisio S, Saiki S, O'Kane CJ, Rubinsztein DC. A rational mechanism for combination treatment of Huntington's disease using lithium and rapamycin. Human Molecular Genetics. 2008;**17**(2):170-178. DOI: 10.1093/hmg/ddm294

[257] Spilman P, Podlutskaya N, Hart MJ, Debnath J, Gorostiza O, Bredesen D, et al. Inhibition of mTOR by rapamycin abolishes cognitive deficits and reduces amyloid-beta levels in a mouse model of Alzheimer's disease. PLoS One. 2010;**5**(4):e9979. DOI: 10.1371/journal.pone.0009979

[258] Wang IF, Tsai KJ, Shen CK. Autophagy activation ameliorates neuronal pathogenesis of FTLD-U mice: A new light for treatment of TARDBP/TDP-43 proteinopathies. Autophagy. 2013;**9**(2):239-240. DOI: 10.4161/auto.22526

[259] DeBosch BJ, Heitmeier MR, Mayer AL, Higgins CB, Crowley JR, Kraft TE, et al. Trehalose inhibits solute carrier 2A (SLC2A) proteins to induce autophagy and prevent hepatic steatosis. Science Signaling. 2016;**9**(416):ra21. DOI: 10.1126/scisignal.aac5472

[260] Aguib Y, Heiseke A, Gilch S, Riemer C, Baier M, Schatzl HM, et al. Autophagy induction by trehalose counteracts cellular prion infection. Autophagy. 2009;**5**(3):361-369

[261] Chen L, Xie Z, Turkson S, Zhuang X. A53T human alpha-synuclein overexpression in transgenic mice induces pervasive mitochondria macroautophagy defects preceding dopamine neuron degeneration. The Journal of Neuroscience. 2015;**35**(3):890-905. DOI: 10.1523/JNEUROSCI.0089-14.2015

[262] Du J, Liang Y, Xu F, Sun B, Wang Z. Trehalose rescues Alzheimer's disease phenotypes in APP/PS1 transgenic mice. The Journal of Pharmacy and Pharmacology. 2013;**65**(12):1753-1756. DOI: 10.1111/jphp.12108

[263] Rodriguez-Navarro JA, Rodriguez L, Casarejos MJ, Solano RM, Gomez A, Perucho J, et al. Trehalose ameliorates dopaminergic and tau pathology in parkin deleted/tau overexpressing mice through autophagy activation. Neurobiology of Disease. 2010;**39**(3):423-438. DOI: 10.1016/j.nbd.2010.05.014

[264] Schaeffer V, Goedert M. Stimulation of autophagy is neuroprotective in a mouse model of human tauopathy. Autophagy. 2012;**8**(11):1686-1687. DOI: 10.4161/auto.21488

[265] Tanji K, Miki Y, Maruyama A, Mimura J, Matsumiya T, Mori F, et al. Trehalose intake induces chaperone molecules along with autophagy in a mouse model of Lewy body disease. Biochemical and Biophysical Research Communications. 2015;**465**(4):746-752. DOI: 10.1016/j.bbrc.2015.08.076

[266] Ledo JH, Azevedo EP, Beckman D, Ribeiro FC, Santos LE, Razolli DS, et al. Cross talk between brain innate immunity and serotonin signaling underlies depressive-like behavior induced by Alzheimer's amyloid-beta oligomers in mice. The Journal of

Neuroscience. 2016;**36**(48):12106-12116. DOI: 10.1523/JNEUROSCI.1269-16.2016

[267] Lourenco MV, Clarke JR, Frozza RL, Bomfim TR, Forny-Germano L, Batista AF, et al. TNF-alpha mediates PKR-dependent memory impairment and brain IRS-1 inhibition induced by Alzheimer's beta-amyloid oligomers in mice and monkeys. Cell Metabolism. 2013;**18**(6):831-843. DOI: 10.1016/j.cmet.2013.11.002